Springer Series in Wireless Technology

Springer Series in Wireless Technology is a series of monographs, contributed titles and advanced textbooks exploring the cutting edge of mobile telecommunication technologies and promulgating them for the benefit of academic researchers, practicing engineers and students. The series encourages contributions in the theoretical, experimental and practical engineering aspects of wireless communications—voice, data and image transmission. Topics of interest to the series include but are not limited to:

- coding and modulation;
- cognitive radio;
- full-duplex wireless communication;
- model-free design;
- multiple access;
- resource allocation;
- uses of digital signal processing in wireless systems;
- wireless energy transfer;
- wireless networks: 4G, 5G and beyond and next-generation WiFi; adhoc wireless networks, device-to-device networks; heterogeneous mobile networks; wireless sensor networks;
- wireless optical communications.

Proposals for this series (please use the proposal form that can be downloaded from this page), can be submitted by e-mail to either the:

Series Editor

Professor **Ramjee Prasad** Department of Business Development and Technology, Aarhus University, Birk Centerpark 15,8001, Innovatorium, CGC, 7400 Herning, Denmark **e-mail:** ramjee@btech.au.dk

or the

In-house Editor

Mr. **Oliver Jackson** Springer London, 4 Crinan Street, London, N1 9XW, United Kingdom **e-mail:** oliver.jackson@springer.com

More information about this series at http://www.springer.com/series/14020

Ramjee Prasad · Vandana Rohokale

Cyber Security: The Lifeline of Information and Communication Technology

Springer

Ramjee Prasad
Department of Business Development
and Technology, CTIF Global Capsule
Aarhus University
Herning, Denmark

Vandana Rohokale
Department of Electronics
and Telecommunication
Sinhgad Institute of Technology
and Science
Pune, India

ISSN 2365-4139 ISSN 2365-4147 (electronic)
Springer Series in Wireless Technology
ISBN 978-3-030-31702-7 ISBN 978-3-030-31703-4 (eBook)
https://doi.org/10.1007/978-3-030-31703-4

This Springer imprint is published by the registered company Springer Nature Switzerland AG
The registered company address is: Gewerbestrasse 11, 6330 Cham, Switzerland

To
My grandchildren
Sneha, Ruchika, Akash, Arya, and Ayush
—Ramjee Prasad

My supportive husband, Milind and
My lovely daughters Madhura and Mugdha
—Vandana Rohokale

Preface

यस्त्विन्द्रियाणि मनसा नियम्यारभतेऽर्जुन।

कर्मेन्द्रियैः कर्मयोगमसक्तः स विशिष्यते॥

Yastvindriyāṇi manasā niyamyārabhate'rjuna|

Karmendriyaiḥ karmayogamasaktaḥsa viśiṣyate||

The person who has control over his/her senses by his/her mind excels in every aspect of life, he/she can perform any task without any attachment with the power of karmyoga.
—The Bhagvad Gita Shloka (3.7)

Cyber security is becoming very hot topic with the huge growth and population of information and communication technology (ICT) and the mobile devices associated with our everyday life. Mobile communication generations from 1G to 4G have changed our lives in many ways. It has brought easiness and comfort in our everyday activities. We are on the verge of welcoming fifth generation of mobile communication that is 5G which is visualized to be converged version of all existing, wired, wireless and next-generation networks. With this luxury, threats are also growing exponentially. Everybody is worried about their economic or intellectual assets. The whole world is looking towards cyber security to provide robust security against the scams or malwares which have penetrated almost everywhere from small sensors to big networks.

Considering the need of in-depth research in cyber security, we have established a strong research group in this field. We decided to write a state of the art on this topic as a book. There are several books available in this field, but none of them has covered broad areas as we have planned to do in this book. This book addresses the cyber security issues starting from cybercrimes to machine-to-machine communication, Internet of things (IoT) and data mining, cyber-physical systems, infected networks called Botnets, E-commerce, social networking, incident handling, smart device security, cloud computing, copyright infringement, artificial intelligence for cyber security and blockchain technology till cyber forensics.

Table 1 Comparison of available cyber security book contents

Sr. no.	Title of the book	Year of publication	Authors	Publisher	Contents
1	Cyber Security: The Lifeline of Information and Communication Technology (ICT)	2019	Ramjee Prasad and Vandana Rohokale	Springer	Various threats and attacks, phishing, M2M communication and IoT, cyber-physical systems, botnet, E-commerce, smart grid security, social networking, incident handling, copyright infringement, fault tolerance, cybercrime, smart device security, AI and ML for next level cyber security, blockchain technology and its usage for providing cyber security, etc.
2	Hacking Exposed 7: Network Security Secrets and Solutions	2018	Stuart McClure, Joe Scambray and George Kurtz	Osborne/McGraw-Hill	Foot-printing, scanning, enumeration, hacking Windows 95/98 and ME, hacking Windows NT, hacking Windows 2000, Novell, NetWare hacking, hacking UNIX, dial-up, PBX, voicemail and VPN hacking, network devices, firewalls, DoS attack, remote control insecurities, advanced techniques, Web hacking, hacking the Internet user, etc.
3	Staying Ahead in the Cyber Security Game	2014	Erik van Ommeren and Marinus Kuivenhoven from the SogetiLabs trend team and Martin Borrett from the IBM Institute for Advanced Security Europe	Capgemini Group and IBM Security	Unequal balance of power between attackers and their victims , the threats on industrial systems, the future of encryption, the implementation of security governance and data protection. Focus on the role data scientists will play moving forward

(continued)

Table 1 (continued)

Sr. no.	Title of the book	Year of publication	Authors	Publisher	Contents
4	Network Science and Cyber Security	2014	Robinson E. Pino	Springer	Intrusion detection systems, behaviour in network traffic, cyber warfare
5	Bio-inspiring Cyber Security and Cloud Services: Trends and Innovations	2014	Aboul Ella Hassanien, Tai-Hoon Kim, Janusz Kacprzyk and Ali Ismail Awad	Springer	Part-I: Bio-inspiring system in cyber security; Part-II: mobile ad hoc networks and key managements; Part-III: biometrics technology and applications; Part-IV: cloud security and data services
6	Counterterrorism and Cybersecurity	2013	Newton Lee	Springer	Counterterrorism in Retrospect: Then and Now, Counterterrorism Technologies: Total Information Awareness and Data Mining, Counterterrorism Technologies: Social Media and Cybersecurity, Counterterrorism Strategies: Causes and Cures, War and Peace
7	The Cyber Index International Security Trends and Realities	2013	James Andrew Lewis and Götz Neuneck	United Nations Publication	Cybersecurity and cyberwarfare: assessment of national doctrine and organization, assessment of international and regional organizations and activities, transparency and confidence-building measures: applicability to the cybersphere
8	Cyber Security Standards, Practices and Industrial Applications: Systems and Methodologies	2012	Junaid Ahmed Zubairi and Athar Mahboob	IGI Global	Mobile and wireless security, social media, botnets and intrusion detection, formal methods and quantum computing, embedded systems and SCADA security, industrial and application security

(continued)

Table 1 (continued)

Sr. no.	Title of the book	Year of publication	Authors	Publisher	Contents
9	Understanding Cybercrime: Phenomena, Challenges and Legal Response	2012	ITU	ITU	The phenomena of cybercrime, challenges of fighting cybercrime, overview of activities of regional and international organizations, anti-cybercrime strategies, legal response
10	Strategic Cyber Security	2011	Kenneth Geers	Kenneth Geers	Cyber security and national security, history, a technical primer, real-world impact, nation state cyber attack mitigation strategies

There are many books available on cyber security with variety of contents. Out of them we have selected only nine books which are more relevant to the subject of our book. As shown in Table 1, the other books have covered various threats, attacks and victims and cyber security for various applications such as biometric technology, cloud security and bio-inspired systems. From the table, it is clear that none of these existing books on cyber security have covered so many diverse topics and issues as compared to our book.

We have tried our best to address the latest topics in this book. The authors are delighted for the diversity of the cyber security issues considered in building this book and the timeliness of these topics. Further suggestions and comments to enhance the book are highly appreciated.

Herning, Denmark Ramjee Prasad
Pune, India Vandana Rohokale

Acknowledgements

Special thanks to Dr. Anand Raghawa Prasad and Dr. Sivabalan Arumugam from Rakuten, Japan, for their visionary inputs towards the completion of the cyber security book. We would like to express our thankfulness to the researchers in the Cyber Security Group for contributing research insights towards diverse cyber security issues.

We are also thankful to the colleagues of GISFI, India, for their untiring efforts. We are falling short of words for the support from the cyber security research team including Supriya, Abhijit, Priyanka, Kirti, Karuna, Javed, Varsha, Swati, Mahesh, Sushma and Jigar for their valuable inputs.

Finally, we would like to acknowledge the support from our family members, CGC, Aarhus University, Vishwaniketan and Sinhgad Institutes.

August 2019

Ramjee Prasad
Vandana Rohokale

Contents

About the Authors

Dr. Ramjee Prasad (Fellow IEEE, IET, IETE and WWRF) is Professor of Future Technologies for Business Ecosystem Innovation (FT4BI) in the Department of Business Development and Technology, Aarhus University, Herning, Denmark. He is Founder President of the CTIF Global Capsule (CGC). He is also Founder Chairman of the Global ICT Standardization Forum for India, established in 2009. GISFI has the purpose of increasing of the collaboration between European, Indian, Japanese, North American and other worldwide standardization activities in the area of information and communication technology (ICT) and related application areas.

He has been honoured by the University of Rome "Tor Vergata", Italy, as Distinguished Professor in the Department of Clinical Sciences and Translational Medicine on 15 March 2016. He is Honorary Professor of University of Cape Town, South Africa, and University of KwaZulu-Natal, South Africa.

He has received Ridderkorset af Dannebrogordenen (Knight of the Dannebrog) in 2010 from the Danish Queen for the internationalization of top-class telecommunication research and education.

He has received several international awards such as: IEEE Communications Society Wireless Communications Technical Committee Recognition Award in 2003 for making contribution in the field of "Personal, Wireless and Mobile Systems and Networks", Telenor's Research Award in 2005 for impressive merits, both academic and organizational within the field of wireless and personal communication, 2014 IEEE AESS

Outstanding Organizational Leadership Award for: "Organizational Leadership in developing and globalizing the CTIF (Center for TeleInFrastruktur) Research Network" and so on.

He has been Project Coordinator of several EC projects, namely MAGNET, MAGNET Beyond, eWALL and so on.

He has published more than 50 books, 1000 plus journal and conference publications, more than 15 patents, over 140 Ph.D. graduates and larger number of masters (over 250). Several of his students are today worldwide telecommunication leaders themselves.

Dr. Vandana Rohokale received her B.E. degree in electronics engineering in 1997 from Pune University, Maharashtra, India. She received her master's degree in electronics in 2007 from Shivaji University, Kolhapur, Maharashtra, India. She has received her Ph.D. degree in wireless communication in 2013 from CTIF, University of Aalborg, Denmark. She is presently working as Professor, in Sinhgad Institute of Technology and Science, Pune, Maharashtra, India. Her teaching experience is around 22 years. She has published one book of international publication. She has published around 35 plus papers in various international journals and conferences. Her research interests include cooperative wireless communications, ad hoc and cognitive networks, physical layer security, digital signal processing, information theoretic security and its applications, and cyber security.

Acronyms

AMI	Advanced Metering Infrastructure
AMPS	Advanced Mobile Phone System
APT	Advanced persistent threat
ARP	Address Resolution Protocol
C and C	Command and control
CASE	Control, Automation and Systems Engineering
CCDCOE	Co-operative Cyber Defense Center of Excellence
CCIE	Control and Industrial Engineering
CD	Cochlear delay
CDMA	Code-division multiple access
CERT	Computer emergency response team
CFS	Correlation-based feature selection
CGH	Computer-generated hologram
CII	Critical Information Infrastructure
COE	Convention on Cybercrime
COMSEC	Communications security
CPM	Cross-platform malware
CPS	Cyber-physical system
CRM	Customer relationship management
CRN	Cognitive radio network
CRTM	Core Root of Trust for Measurement
CSE	Consistency-based subset evaluation
CSIRC	Computer Security Incident Response Capability
CSIRT	Computer security incident response team
DCT	Discrete cosine transform
DDOS	Distributed denial of service
DFRWS	Digital Forensic Research Workshop
DFT	Discrete Fourier transform
DITSO	Defense Information Technology Services Organization
DLP	Data loss prevention

DR	Demand response
DWT	Discrete wavelet transform
DYWT	Dimensional Dyadic Wavelet Transform
ECC	Elliptic-curve cryptography
ERP	Enterprise resource planning
ETCM	Ecuador Technical Chapters Meeting
FDI	False data injection
FDMA	Frequency-division multiple access
FFT	Fast Fourier transformation
FTM	Fault tolerance manager
GAN	Generative adversarial network
GIMCV	Global Information Multimedia Communication Village
GMSK	Gaussian Minimum Shift Keying
GOZ	Gameover ZeuS
GP	Genetic programming
GPRS	General Packet Radio Service
GSM	Global System for Mobile Communications
HAN	Home area network
HIL	Hardware in the loop
HVS	Human visual system
IAAS	Infrastructure as a service
ICANN	Internet Corporation for Assigned Names and Numbers
IBC	Identity-based cryptography
IBG	Industry Botnet Group
ICEOE	International Conference on Electronics and Optoelectronics
ICMP	Internet Control Message Protocol
ICMT	International Conference on Multimedia Technology
ICNSC	International Conference on Networking, Sensing and Control
ICRCC	International Conference on Radar, Communication and Computing
ICT	Information and communications technology
IDAACS	International Conference on Intelligent Data Acquisition and Advanced Computing System
IDS	Intrusion detection system
IJCSE	International Journal on Computer Science and Engineering
IOCE	International Organization on Digital Evidence
IOT	Internet of things
IPR	Intellectual property rights
IRC	Internet relay control
IRS	Internal Revenue Service
IRT	Incident response team
ISCIT	International Symposium on Communications and Information Technologies
ISIRT	Information Security Incident Response Team
ISM	Industrial, scientific and medical
ISSA	IEEE Information Security for South Africa

ISSRE	International Symposium on Software Reliability Engineering
IT	Information technology
ITS	Intelligent transport system
ITU	International Telecommunication Union
LALR	Look-ahead left to right
LAN	Local area network
LE	Law Enforcement Investigative
Li-Fi	Light fidelity
LLFT	Low Latency Fault Tolerance
LLP	Lower-level problem
LPWAN	Low-power wide-area network
LSB	Least significant bit
MAB	Multi-armed bandit
MAC	Media access control
MIC	Measurement, Information and Control
MITM	Man-in-the-middle attack
MoWNeT	Mobile and Wireless Networking
MPI	Message Passing Interface
MTC	Machine-type communication
NAN	Neighbourhood area network
NDSS	Network and Distributed System Security Symposium
NFC	Near-field communication
NIDS	Network intrusion detection system
NIST	National Institute of Standards and Technology
NMT	Nordic Mobile Telephone
NOMS	Network Operations and Management Symposium
NR	Network reliability
nslookup	Name Server Lookup
OAuth	Open Authentication
OIDC	OpenID Connect
OTPS	One-time passwords
P2P	Peer-to-peer
PAAS	Platform as a service
PCA	Principal Component Analysis
PCFG	Probabilistic Context-Free Grammar
PIN	Personal identification number
PLM	Product lifecycle management
QOE	Quality of Experience
QOL	Quality of life
QOS	Quality of service
RAdAC	Risk Adaptable Access Control
RAT	Remote access Trojan
RCFL	Regional Computer Forensics Laboratory
RFID	Radio-frequency identification
RTR	Root of Trust for Reporting

RTS	Root of Trust for Storage
SAAS	Software as a service
SAML	Security Assertion Markup Language
SDN	Software-defined networking
SDR	Software-defined radio
SEM	State estimation model
SFD	Self-tuning failure detector
SIEM	Security information and event management system
SIG	Special Interest Group
SMI	Smart metering infrastructure
SOP	Standard operating procedure
SV	Sampled value
SVM	Supervised Machine Learning
SWGDE	Scientific Working Group on Digital Evidence
TC	Trusted Computing
TCG	Trusted Computing Group
TPM	Trusted Platform Module
ULP	Upper-level problem
VANET	Vehicular ad hoc network
VBN	Virtual Business Portals
VBR	Volume boot record
VFT	Virtualization and Fault Tolerance
VLAN	Virtual local area network
VLC	Visible light communication
VM	Virtual machine
WAN	Wide-area network
WISDOM	Wireless Innovative System for Dynamically Operating Mega-communications
WSN	Wireless sensor network

List of Figures

List of Tables

Chapter 1
Introduction

Digitization is becoming the basis of future development in our society and economy. In today's world, People, devices and machines are networked through wired or wireless means. Now is the era of Internet and smart mobile devices. Internet has touched every human being and has changed the way we perform our everyday activities like working, playing, shopping, seeing movies and serials, talking on the phone, listening to our favorite music, ordering food, paying bills, making friends, and greeting our friends and relatives on their special occasions. Due to this every time and everywhere connectivity, track of every user and the objects is possible using the IP address. At this point, users can not stop using Internet but they expect it to be secure, privacy preserving and trustworthy.

Cyber security issues happening not due to a magical wand by attackers but due the vulnerability present in the system gets exploited. Presently we are living in a world where all of us are surrounded by numerous heterogeneous wireless devices and technologies. Attack on Information Technology (IT) network is one thing; but attack on mobile networks is a matter of life and death especially with mobile networks being the critical information infrastructure (CII).

As we move ahead we will see television white spaces and other spectrum white spaces are being utilized for emerging spectrum need for the broadband Internet access. Together with that fifth generation of mobile communication system (5G) is knocking our door in the form of user-centric system. 5G will be an extended version of Global Information Multimedia Communication Village (GIMCV) and Wireless Innovative System for Dynamically Operating Mega-communications (WISDOM) concepts envisioned around 2008 (Prasad 2008). Moving from mobile networks being Critical Information Infrastructure (CII) today to 5G tomorrow (in near future), cyber security attacks will have implications that can go beyond imagination. This chapter is structured as follows: Sect. 1.1 explains the emerging cyber threats. Cyber Security is discussed in Sect. 1.2. Section 1.3 explains the correlation between mobile communication and cyber security. Purpose and structure of the book are discussed in Sect. 1.4. Finally, Sect. 1.5 summarizes the introduction chapter.

R. Prasad and V. Rohokale, *Cyber Security: The Lifeline of Information and Communication Technology*, Springer Series in Wireless Technology,
https://doi.org/10.1007/978-3-030-31703-4_1

1.1 Emerging Cyber Threats

Cyber security is the assemblage of skills, techniques, processes and run-throughs which are built for ensuring the protection of the network, computers, programs and data against malware, attacks, damage and unauthorized access. Mobile communications has brought ease and luxury in our lives. The progress and developments in Information and Communications Technology (ICT) in the form of mobile communications towards 5G is no doubt inevitable. But the negative side in the form of threats to our assets makes us insecure. ICT systems are hacked every day for robbing money and business secrets, for political aspirations or for stealing intellectual property.

Mat Honan, a senior staff writer with Wired, had been a victim of cyber attack in 2012. He had explained in details, how within one hour of time span, the attackers had erased all of the data of his iPhone, iPad, and MacBook by getting illegal access to his Twitter and Google accounts (Honan 2012). The attackers also deleted Honan's Google account, including almost eight years worth messages from his Gmail inbox. This attack compromised Honan's information on all three levels that is, confidentiality, integrity, and authenticity (C–I–A) model of information security.

Figure 1.1 depicts different vulnerabilities surrounding us every day. There are emerging number of vulnerabilities (Shah and Ravi 2018) including spyware, virus,

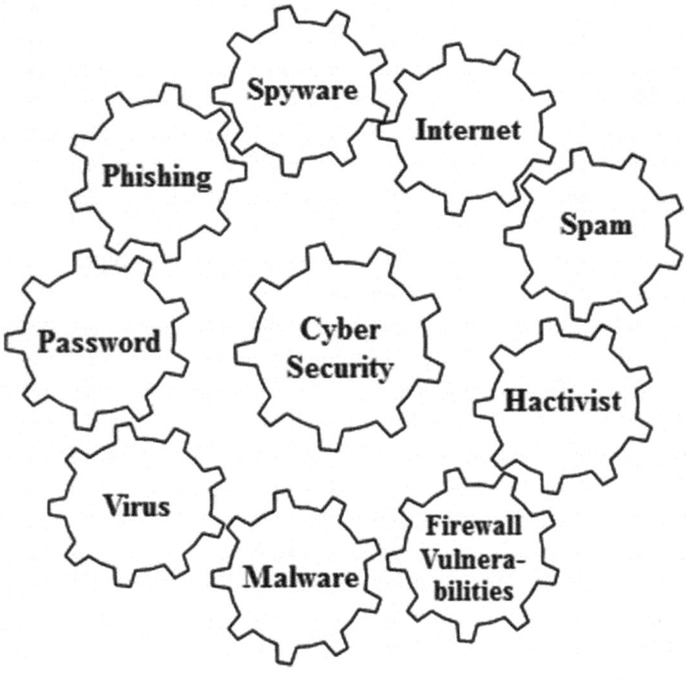

Fig. 1.1 Emerging cyber security threats

malware, phishing, hacking, spam, Internet worm, identity thefts, password scams, firewall vulnerabilities, etc. Cyberterrorism is the intentional use of Information and Communication Technology devices and techniques for malign purposes. Computers, mobile devices, networks and Internet can be used for destructive causes for personal gains or political benefits. Cyberwar and cyberterrorism go hand in hand with each other. By making use of cyber tools, cyber physical infrastructure such as finance, energy, transport and government are targeted by cyber attacks to increase terror and physical injuries or deaths (Ivanov 2019). New cyber threats are emerging everyday including ransomware, endpoint attacks, phishing, third party attacks, supply chain attacks, artificial intelligence and machine learning driven attacks, cryptojacking, cyber physical attacks, state sponsored attacks, IoT attacks, threats to smart medical devices and electronics medical records, attacks on connected cars, semi-autonomous vehicles and driverless cars (O'Brien 2018). The whole world is in need of strong cyber security techniques which can successfully combat these ever increasing forms of threats. Next section elaborates on cyber security and its importance in the world of digitization.

1.2 Cyber Security

Global digitization brings huge opportunities and promises ease of life. Smart Cities is the first and very essential step towards it. It promises affordable housing for economically backward people, efficient urban mobility and public transport, adequate clean water supply, assured electricity supply, sanitation, robust IT connectivity, good governance with citizen participation, sustainable environment, health and education facilities, and last but not the list safety and security of citizens. In proportion to this growth, the risk of cyber-attacks on the ICT network infrastructure which are used for smart city are also increasing (Rohokale and Prasad 2017).

As per ITU-T definition, "Cyber security is the collection of tools, policies, security concepts, security safeguards, guidelines, risk management approaches, actions, training, best practices, assurance and technologies that can be used to protect the cyber environment and organization and user's assets. Organization and user's assets include connected computing devices, personnel, infrastructure, applications, services, telecommunications systems, and the totality of transmitted and/or stored information in the cyber environment. Cybersecurity strives to ensure the attainment and maintenance of the security properties of the organization and user's assets against relevant security risks in the cyber environment" (ITU-T 2017).

Today personal and business life is completely dependent on interconnected things and the automation in almost every aspect of human life which has brought smartness and luxury into everyday life. These smart devices generate huge data which may include some critical and sensitive information that is stored on the cloud. This data is vulnerable to variety of attacks intending data thefts and/or data leakage. ICT counterparts need to be more ready for these kind of threats and vulnerabilities. Security is not a one-time course of action; it is a continuous process which needs innovative defense mechanisms. Next section discusses the security provisions in the mobile communication networks and also discusses the cyber security scenario in nextgen mobile communication network that is 5G.

1.3 Mobile Communication and Cyber Security

Mobile communications is life-line of the society today. As mobile communications evolves, it is becoming prey of cyber security as well. This section explains the history of mobile communication technologies.. The first generation of mobile communication was started in the 1980s. First generation (1G) mobile phones were analog in nature with frequency division multiple access (FDMA) techniques employed for spectrum access. The demand for radio spectrum in 1G was huge due to FDMA technique. Deployment of first analog mobile phones in Nordic European countries like Finland, Sweden, Denmark and Norway took place in 1981. Two versions of Nordic Mobile Telephone (NMT) with frequency bands of 450 and 900 MHz named as NMT-450 and NMT-900.

Advanced Mobile Phone System (AMPS) was the first analog cellular system commercially deployed in North America and Israel during 1983 and 1986 respectively. AMPS were the real cellular system which was accepted by mass user market, but it is insecure due to lack of appropriate security mechanism. It was prone to eavesdropping attack and also vulnerable to cloning attack (Farley and Van Der Hoek 2006).

Second generation (2G) Digital Mobile Phone System emerged around the 1990s. It followed two competent standards namely Global System for Mobile (GSM) and Code Division Multiple Access (CDMA) was introduced by Europe and United States respectively. The 2G mobile phones has the ability to access media content through the Internet on a portable screen (MMS). The 2G digital mobile system were able to provide data rates ranging from 64 to 144 kbps. GSM, TDMA (IS-136) and CDMAone (IS-95A) were the standards evolved for 2G. The 2.5G mobile system is the interface between 2G and 3G. It provides additional features such as packet switched connection General packet radio service (GPRS) and 2.75G provides Enhanced Data rates for GSM Evolution (EDGE). The GPRS comes under the umbrella of GSM whereas the EDGE is used by both the GSM and the CDMAone. The 2.5G networks provide much higher data rates ranging from 40 to 236 kbps and the 2.75G offers data rates up to 384 kbps. The modulation technique used in 2G system was *Gaussian Minimum Shift Keying* (GMSK). It provides services such as multimedia messaging and Internet access.

In 2000, International Telecommunication Union (ITU) has set a standard for 3G called International Mobile Telecommunication-2000 (IMT-2000). Due to the seamless connectivity feature provided by 3G system, the world truly got shrunk into a small connected village. The 3G mobile communication offers applications such as global roaming, wireless voice telephony, mobile Internet access, fixed wireless Internet, video telephony, and mobile TV. The 3G technologies support data rates up to 2 Mbps. Voice calls are connected through circuit switching, but the multimedia data are sent through packet switching (Dholakiya and Jain 2001; Prasad et al. 2000).

Wireless Digital Broadband Internet access on mobile phones is possible due to fourth generation (4G) mobile communication standard. In December 2010, ITU approved IMT-Advanced as official 4G standards i.e. SAE/LTE (Mshvidobadze 2012).

Whooping speed is the most attractive feature of 4G. It is considered as a mobile multimedia with anytime anywhere connectivity and global mobility support and customized personal service network system with a integrated wireless solution (Kim and Prasad 2005). Some of the important features of 4G include all-IP packet switched network, high data rates up to 1 Gbps, Seamless connectivity, and global roaming, Interoperability with existing standards, Smooth handovers and high QoS (Prasad and Munoz 2003; Glisic and Lorenzo 2009). The authors in (Prasad et al. 2009) have visualized the shrieked world with 4G as Global Information Multimedia Communication Village (GIMCV) for which the cellular cells are moving from global to pico-cellular size. The necessity of collaborative functionality of WLANs and WPANs in coordination with 3G systems is reasonable with GIMCV paving the way for 4G. The key issues in 4G networks are strongly addressed in the literature related to advance wireless networks (Glisic and Lorenzo 2009; Prasad et al. 2009). It throws light on cognitive, cooperative and opportunistic approaches which are capable of advancing network efficiency further.

Issues related to almost all emerging networks like wireless Internet, mobile cellular, WLAN, WSN, the Adhoc networks, bio-inspired networks, active and cognitive networks with cooperative strategies which are present in the 4G due its flat IP network nature.

The majority of economic growth in today's wireless world is driven by Information and Communication Technology (ICT). Society and digital economy are closely interconnected with each other via critical infrastructures like energy and electricity, water, food, transportation, public health, telecommunication and ICT, different government services, etc. The present generation networks are either service oriented or operator-centric. But the nextgen network revolution is going to be fully user centric with better quality of experience (QoE) for the user. Table 1.1 shows the parametric comparison of different generation of mobile communications along with its security features.

The converged visualization of 5G mobile wireless is depicted in Fig. 1.2. Converged 5G mobile wireless system is an integration of different network technologies like wired and wireless, old and newly evolved networks and supporting technologies. Wireless Sensor Networks (WSNs), Internet of Things (IOT), WLAN, Wi-Fi, Light Fidelity (Li-Fi) with Visible Light Communication (VLC), Cellular Networks, Infra-structured networks like PSTN, Cognitive Radio Networks (CRN) with Software Defined Radios (SDR) and Software Defined Networks (SDN), Vehicular Adhoc Networks (VANETs), Smart Grids, Optical Fiber Communication, Cloud Computing, Tele-healthcare systems, cooperative

Table 1.1 Parametric comparison of 1G–5G

Mobile communication generation/ parameters	1G	2G	2.5G	3G	3.5G	4G	5G
Deployment	1980	1990	1999	2000	2010	2010+	2020
Analog/digital	Analog	Digital	Digital narrowband	Digital wideband	Digital wideband	Digital broadband	Converged
Architecture	Circuit switched	Circuit switched	Packet switched	Circuit and packet switched	Packet data	All IP	Mobile IP
Core network	PSTN	PSTN	PSTN, packet network	Packet network	Internet	Broadband internet	Broadband internet
Access mechanism	FDMA	TDMA CDMAone (IS-95A)	TDMA CDMAone (IS-95B)	TDSCDMA CDMA2000	OFDMA CDMA	OFDMA SCFDMA	LASCDMA BDMA, WDM
Standard	AMPS, TACS, NMT	GSM, TDMA	GPRS, EDGE	WCDMA (UMTS), EVDO	HSPDA, HSPA+	LTE, LTE-A, IMT-advanced, WiMAX2	CRN, SDR, SDN, NFV
Services provision	Voice	Voice, SMS	Voice, SMS, MMS	Voice, text, multimedia	High-speed internet	Multimedia, video streaming, virtual gaming	User centric high-speed converged applications
Coverage	Outdoor	Indoor/ outdoor	Partially global	Global roaming	Global roaming	Global roaming	Global roaming
Data rates	2–14.4 kbps	14.4–64 kbps	64–144 kbps	384–2 Mbps	14–56 Mbps	100–1 Gbps	1–1 Tbps
Security provision	Insecure due to lack of encryption	Secure with symmetric key algorithm	GPRS encryption algorithm (GEA)	Wireless application protocol WAP1, WAP2 and WIM	AES, IPsec, AKA	Authentication and key agreement (AKA), IPsec, IKE, GBA-IdM	Network coded physical layer, link layer, dynamic security

Fig. 1.2 Visualization of 5G network with cyber crimes and security provision

multiuser MIMO with smart antennas, millimeter wave communications, satellite communications, etc. will have to work in collaboration and cognition with every other network to bring 5G into reality. The user-centric approach of 5G, Quality of Life (QoL), Quality of Experience (QoE) by user and quality of Service (QoS) by service providers will have to walk hand in hand.

5G was visualized as a mobile wireless system ahead of time around 2010 through GIMCV and Wireless Innovative System for Dynamic Operating Mega Communications (WISDOM) (Prasad 2012). Cognitive radio networks and personal area are integrated together to achieve robustness and efficiency in the future wireless networks (Prasad 2014). WISDOM proposed a smooth shift from static to dynamic communication and ensures secure user experience with the help of virtual business portals (VBNs). It envisioned the seamless connectivity for users by bridging virtual and physical worlds with sensitive and context based amusing the user experience over wired and wireless networks. It focused on the bandwidth hungry applications including entertainment, multimedia, Intelligent Transport Systems (ITS), Telemedicine, emergency and safety applications. WISDOM offered the data rates up to 1 Tbps. Figure 1.3 depicts the objectives of the WISDOM. The 5G is envisioned as the combination of 4G and WISDOM.

According to the author in (Prasad 2014), "WISDOM is a communication system for ubiquitous, reliable human-centric connectivity via an arbitrary

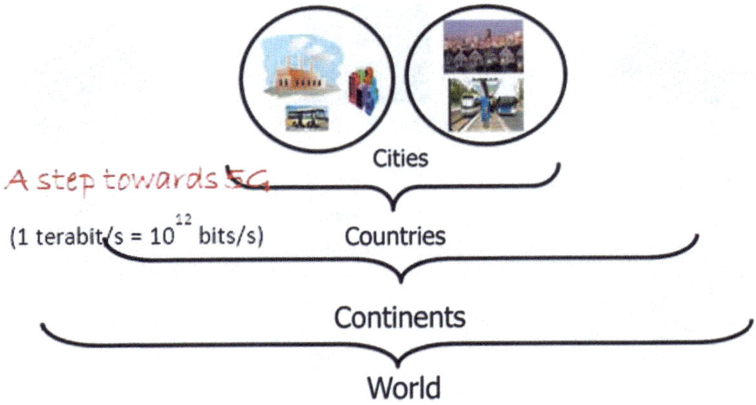

Fig. 1.3 WISDOM concepts (Prasad et al. 2009)

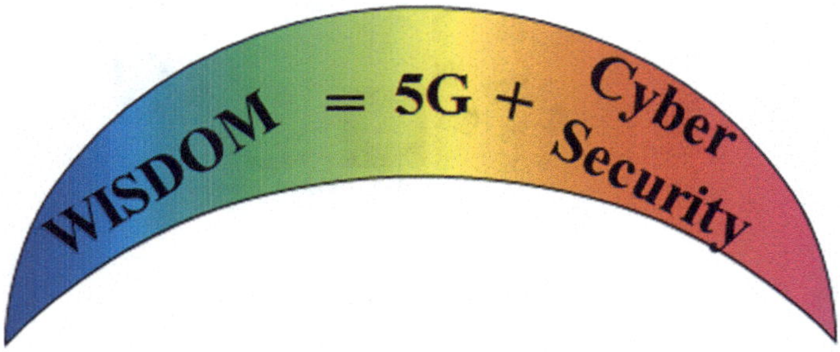

Fig. 1.4 5G with provision of cyber security is WISDOM

infrastructure support". WISDOM ensures security by design with the end to end security provisions. Next generation networking technologies should be able to provide persistent trust and privacy in the usage of ICT. If upcoming 5G networks can offer security with the converged, fast and energy efficient communication, then WISDOM will come into real practice. Figure 1.4 indicates the WISDOM concept as the integration of 5G and cyber security. In the real practical sense, cyber security is going to be the lifeline of tomorrow's green and secure ICT.

1.4 Purpose and Structure of the Book

This book touches almost all aspects of cyber security and presents its importance for mobile communication system. Figure 1.5 elaborates different issues and aspects of cyber security which are considered as the individual chapters in the book.

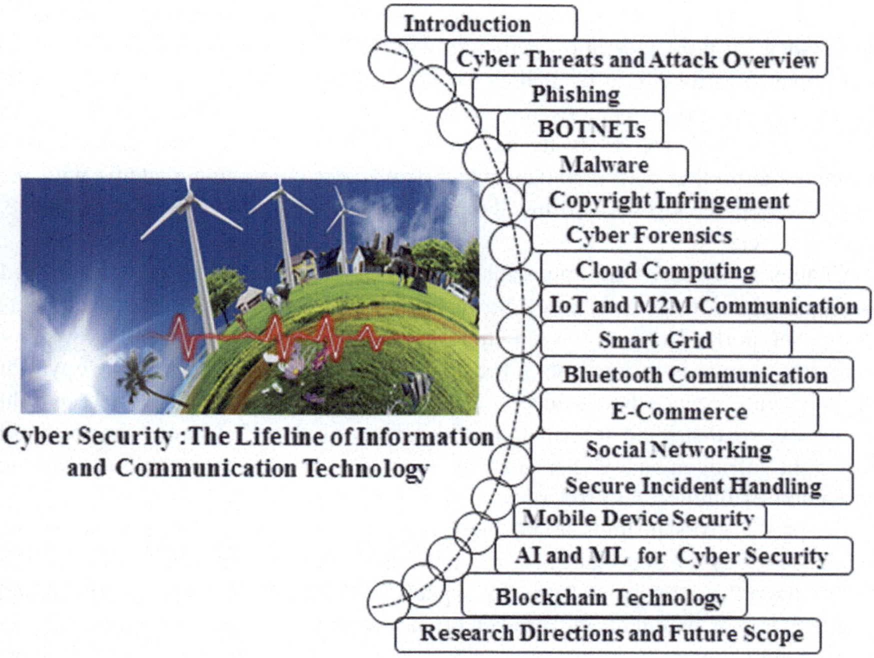

Fig. 1.5 Organization of the book with different cyber security issues

Chapter 2 discusses the overview of cyber threats and attacks. Most of the known cyber-attack techniques include advanced persistent threat (APT), distributed denial of service (DDoS), cross-platform malware (CPM), metamorphic and polymorphic malware and phishing are discussed in this chapter. New cyber-attacks including Ransomware, endpoint attacks, phishing, third party or supply chain attacks, artificial intelligence and machine learning driven attacks, cryprojacking, cyber physical attacks, state sponsored attacks, IoT attacks, threats to smart medical devices and electronics medical records, attacks on connected cars, semi-autonomous vehicles and driverless cars are also elaborated in this chapter.

Chapter 3 explains the need for cyber security in phishing scenarios. Personal details of the online user are usually obtained by cheating them with the help of emails, advertisements and some sites which are usually browsed by the online users. After the occurrence of incidences, users have to face other information robberies. Secure communication is the right of every user in the cyber wireless world.

Chapter 4 gives a thorough idea about the infected networks and importance of cyber security in such kind of large variety of infected networks called BOTNETs. A BOTNET is a collection or a network of infectious 'bots' i.e. machines. There is variety of botnets. Botnets have become a platform for the infection to the Internet, such as, spam, e-mails, launch denial of services attacks, click fraud, cyber warfare, cyber sabotage, etc. Detection of such BOTNETs and repairing is of great importance for next-generation networks with M2M communication.

Chapter 5 throws light on malicious software that is Malware and the investigation and deception methods for them. It also discusses malware analysis and virtualization malware. Malware can be in the form of quick executable codes, scripts or any other active content which can infect computing systems with the help of computer viruses, worms, Trojan horses, Ransomware, spyware, adware, shareware, etc.

Chapter 6 discusses the important aspect of Intellectual Property Rights named as Copyright Infringement case. Nowadays, protecting the electronic documents from unauthorized copying is an important issue. Due to this rapid growth of technology, there is an increase in the different cyber-crime attacks risk. Copyright infringement occurs when someone other than the copyright holder or copyright owner use or distribute information without permission. It is not easy to avoid copyright infringement. Cyber security must provide robust techniques for the protection of Intellectual properties.

Chapter 7 gives insights about the Cyber Forensics field. While handling the incidences, it is an important task to gather the evidence, analyze them, preserve and manage the security and privacy aspects of the users. It is a very critical and sensitive job to maintain and protect privacy without hurting the individuals by taking help of their electronic assets. Chapter 8 highlights the importance of Trust Management in whole Cyber Security aspects. Information assurance factors such as authenticity, authorization, non-repudiation, confidentiality and integrity of data are of utmost importance which has to be maintained in digital communication. Trust management is one of the mutual approaches where any of the peers from the network can assure its security from the malicious attack. Several trust management schemes have been proposed recently to counter the security threat on Peer-to-Peer (P2P) systems.

Chapter 8 elaborates cyber security in the Cloud Computing world. Almost all the data from interconnected systems is stored in the cloud. So, cloud security is of most importance. Fault tolerance is the most critical issue in cloud computing platforms. Most cloud computing platforms use virtualization. Due to this, the cloud computing platform is splitted into three main layers like hosts, virtual machines and applications. In the view of administrative side, there are two layers such as cloud provider and customer.

Chapter 9 deals with Internet of Things (IoT) and Machine to Machine Communication. With 5G, the mobile services are going to get transformed from Cell Centric approach to User Centric Approach. Till 4G, people were communicating with each other with the aid of electronic gadgets, but next generation of mobile communication is going to witness the communication between intelligent objects making use of Radio Frequency Identification (RFID) and Sensors. These ubiquitous networks are going to produce big data and numerous threats. Cyber Security is crucial for such networks.

Chapter 10 throws light on the security of Smart Grids. Electricity thefts, wastage, and appropriate usage management are the key points under consideration.

Cyber Security is going to play the vital role in making power generation, distribution and management aspects Green and Secure. Advanced Metering Infrastructure (AMI) is suffering from a lot of security threats. Remote control and monitoring make these systems more vulnerable to cyber-attacks, so robust security measures are essential for the persistent life of cyber-physical systems.

Chapter 11 elaborates Bluetooth Enabled Communication and cyber security issues, attacks and available security algorithms. Probable security solutions for this are discussed in the chapter. Bluetooth technology is an emerging wireless networking standard, which is based on chip providing short-range wireless communication. However, data-leaking is one of the serious and frequently arising phenomena in the use of Bluetooth technology for data transfer. Some Bluetooth enabled communication-related attacks with possible security solutions are deliberated here.

Chapter 12 gives an idea about cyber security aspect for E-Commerce or Online Business. As the Internet is extensively used, protection of information during online transactions has been a very significant aspect as far as national, and personal asset security is concerned. The craze for online shopping is increasing every day. Although it relieves human efforts, time, and energy, it is more prone to cyber threats, and lots of attacks are possible. It is natural user expectation to have their online transactions to be secure, and freedom of privacy should be maintained. Cyber Security is going to prove a robust solution for e-commerce business transactions.

Chapter 13 deals with cyber security for Social Networking communications like WhatsApp, Facebook, Twitter, etc. Social Networks apparently have become a part of most literate human beings. As compared to the time they spend for other activities, for the majority of the time, they stick with the Facebook, Twitter, LinkedIn, etc. The increase in the number of users has amended the hackers to steal or get access to other people's social network profile and use it for some illegal activities. The malicious code injected in the web through different kind of sources like making the users view eye catching flash ads, redirecting the users to malicious content pages, downloading applications along with malicious code and through a lot of ways the viruses, worms are spread into the web with a purpose of malicious activities involved. Security provision for such kind of fascinating social networking sites is very fragile. Cyber security must ensure these communications to be private and secure.

Chapter 14 elaborates secure incident handling. Appropriate advanced security planning makes sure that all the necessary response processes are well known, well-coordinated and need to be carried out steadily. This chapter includes secure incident handling processes such as preparation, detection and identification, suppression, abolition, recovery, and consequences. Chapter 15 gives idea about the mobile device cyber security. Mobile devices now a day provides a lot of facilities such as social network, banking, online shopping, games etc. Since this device is so useful to us so we need to take care of the security issues related to it. Mobile device security is the full protection of data on portable device and the network connected to the device.

Chapter 16 illustrates how artificial intelligence and Machine learning can be exploited in attainment of cyber security. By making use of advanced machine learning and artificial intelligence techniques, one step forward security solutions can be obtained. Blockchain Technology and its use for achieving cyber security are discussed in Chap. 17. Cryptography is applied to prepare an ever-growing list of digital records called as Blockchain. Each block is the cryptographic hash of the previous block, a timestamp and data to be transacted. It is immune to the changes in data once it is a part of the Blockchain digital ledger. Because of its virtual nature, Blockchain data becomes almost incorruptible. Concluding Chap. 18 throws light on some imperative research challenges and future scope of the research work. This chapter presents new research directions and thoughts to the emerging researchers who are interested in the cyber security field.

1.5 Summary

The mobile communication industry is marching towards 5G with full pace. For converged 5G wireless networks and services, cyber security is a crucial issue. Mobile communication of future that is 5G will be an integrated version of Internet of things and machine to machine communication, cloud computing, smart grid, smart devices, etc. So it is very much timely and essential to consider security issues for upcoming and existing technologies. Cyber security is an integral part of the future information and communication technology designs and implementations.

References

Dholakiya JH, Jain VK (2001) Technologies for 3G wireless communications. In: International conference on information technology, coding and computing, pp 162–166

Farley T, Van Der Hoek M (2006) Cellular telephone basics. Article posted on Privateline

Glisic S, Lorenzo B (2009) Advanced wireless networks: cognitive, cooperative and opportunistic 4G technology, 2nd edn. Wiley Publications

Honan M (2012) Wired magazine, gear. How apple and amazon security flaws led to my epic hacking. 8th June 2012. https://www.wired.com/2012/08/apple-amazon-mat-honan-hacking/

ITU-T (2017) ITU-T: committed to connecting the world, study group 17. Cybersecurity. https://www.itu.int/en/ITU-T/studygroups/com17/Pages/cybersecurity.aspx

Ivanov I (2019) Cyber security and cyber threats: eagle vs 'new wars'? Academia Publishing

Kim Y-K, Prasad R (2005) 4G roadmap and emerging communication technologies. Artech House Publishers

Mshvidobadze T (2012) Evolution mobil wireless communication and LTE networks. In: 6th international conference on application of information and communication technologies (AICT), pp 1–7

O'Brien D (2018) The A to Z of cyber security. Symantech. https://medium.com/threat-intel/the-a-to-z-of-cyber-security-93150c4f336c

Prasad R (2008) Keynote speech—wireless innovative system dynamic mega communications (WISDOM). In: First IEEE international workshop on cognitive radio and advanced spectrum IEEE CogArt'08

Prasad R (2012) Future networks and technologies supporting innovative communications. In: Keynote speech in proceedings of IC-NIDC

Prasad R (2014) 5G: 2020 and beyond. River Publisher Series in Communications

Prasad R, Munoz L (2003) WLANs and WPANs towards 4G wireless. The Artech House Universal Personal Communications, Boston, London

Prasad R, Mohr W, Konhauser W (2000) Third generation mobile communication systems. Artech House Publishers

Prasad R, Pruthi P, Ramareddy K (2009) The top 10 list for terabit speed wireless personal services. J Wirel Pers Commun. (Special issue on GIMCV)

Rohokale V, Prasad R (2017) On-going and future breakthroughs in smart city developments. In: Ligthart L, Prasad R (eds) Role and importance of the cyber security for developing smart cities in India. River Publishers, pp 117–136

Shah A, Ravi S (2018) A to Z of cyber crime. Asian School of Cyber Laws, Lexcode Education and Assessment Platform (LEAP)

Chapter 2
Cyber Threats and Attack Overview

For cyber security, risk is the integrated effect of vulnerabilities, threats, and potential impact of cyber-attacks. Vulnerability is the potential weaknesses in the cyber security system. Threat is a possibility of cyber-attack by making use of system vulnerabilities. Internet and Internet of things (IoT) are the major threat entities. Data threat is increasing with scaling of new web applications. Some plug-ins are letting malwares enter in the system. Phishing is getting smart and passwords no longer guarantee security. Main threat actors are cyber criminals, nation states, and hacktivists.

Some of the recent biggest cyber-attacks include operation shady RAT on US government, large scale financial hack attack on American firm TJX, attack on New Jersey's Heartland Payment System, credit card data hack through attack on email marketing firm Epsilon, attack on Sony play station in which credit card data for more than 70 million people was compromised (The Telegraph, Internet Security 2017). This chapter is organized as follows. Section 2.1 discusses the categorization of cyber-attacks. Typical attack sequence is explained in Sect. 2.2. Section 2.3 elaborates different types of cyber-attacks. Footprinting attack is described in Sect. 2.4. Section 2.5 covers wiretapping attack. Social engineering attack is explained in Sect. 2.6. Section 2.7 discusses packet sniffing attack. Well known ports and port scanning attack are described in Sect. 2.8. Next Sect. 2.9 throws light on password vulnerabilities. Section 2.10 talks about track covering. Malwares are narrated in Sect. 2.11. Section 2.12 explains viruses and worms in details. Information about logic bomb attacks is covered in Sect. 2.13. BOT and BOTNET are elaborated in Sect. 2.14. Last Sect. 2.15 gives idea about Trojan horse attack.

R. Prasad and V. Rohokale, *Cyber Security: The Lifeline of Information and Communication Technology*, Springer Series in Wireless Technology,
https://doi.org/10.1007/978-3-030-31703-4_2

2.1 Cyber Attack Categorization

The continuous advent of new targeted attacks with advanced persistent threats demands the development of powerful cyber security techniques with new approaches. The cyber-attack surveys indicate that almost every aspect of human life is suffering from cyber-crimes. From energy services to financial, manufacturing, public, travel, health, retail, professional services are seen to be victims of one or the other cyber-crime.

The number and severity of the cyber-crimes are ever increasing with the advent of the new mobile computing technologies and handheld gadgets. In these cases, the computing systems may have been used to initiate the attack or the same system can be the victim of such attacks (Khanse 2014). Various categories of cyber-attacks include malware, network attacks, network intrusion attacks, social engineering attacks, cyber espionage, reconnaissance, network access attacks, cyber terrorism, cyber warfare, etc., as shown in Table 2.1. Next section describes typical attack sequence during a cyber attack.

Table 2.1 Cyber attack categorization

Sr. no.	Attack category	Attack description	Sub-attacks
1	Malware	Malicious software used to launch specific attacks in the computer systems	Adware, Spyware, Virus, Worm, Trojan, Rootkit, Backdoors, Key loggers, Rogue Security Software, Ransomware, Browser Hijacker, etc
2	Network attack	Active or passive monitoring of computer communications and network traffic	Passive, Active, Distributed, Insider, Close-in, Phishing, Hijacking, Spoofing, Buffer overflow, Exploit, Password attack, etc.
3	Network intrusion attacks	Any unauthorized activity on the computer networks	Asymmetric routing, Buffer overflow, Protocol specific attacks, Traffic flooding, Trojans, Worms, etc.
4	Social engineering attacks	Using social media and phone calls, attackers apply human psychology trick to make users giving access to sensitive information	Phishing, Pre-texting, Baiting, Quid Pro Quo, Tailgating, etc.
5	Cyber espionage	Snooping on confidential information of a user or organization without permission	Industrial espionage, Nation-State espionage, Economic espionage, Corporate espionage, Information theft and sabotage

(continued)

Table 2.1 (continued)

Sr. no.	Attack category	Attack description	Sub-attacks
6	Reconnaissance	By finding out weaknesses in the network systems and services, attacker gathers sensitive information about the network	Internet information lookup, Ping sweeps, Port scans, Packet sniffers
7	Network access attacks	By searching out malicious activities in the network authentication, FTP and web services, the intruder gets access to a network system to obtain confidential information	Eavesdropping, Data modification, Identity spoofing, Password-based, Denial of service, man in the middle attack, Compromised key attack, Sniffer attack, Application layer Attack, Trust exploitation, etc.
8	Cyber terrorism	Use of internet for electronic terrorist activities like large-scale disruption of computer networks, high-profile national components, national critical infrastructures or important business operations	Sabotage, Website defacement and Denial of service, Destruction of critical physical infrastructures, etc.
9	Cyber warfare	Major disruption to national critical and highly important infrastructures through malign use of digital information	Disruption of nation's public services, Financial Institutions, Industrial espionage, blocking military and civilian responders, etc.

2.2 Typical Attack Sequence

With ever growing cyber-crimes and their sternness, it is necessary to understand the steps cyber criminals take to attack user's computing systems so that appropriate cyber security mechanisms can be developed to combat these cyber-crimes. The basic steps cyber criminals follow during cyber-attack include finding vulnerabilities in the target system, which is referred to as reconnaissance, then actual penetration into the network through intrusion followed by insertion of malicious secret codes into target system referred to as malware and the last step is to clear the visited tracks which is referred to as clean up stage (Identity Week 2017; de Ramos blog 2016). Typical cyber attack sequence is as follows:

(1) Reconnaissance and Scanning: The attackers first study the system in details to find the weaknesses and then by exploiting the vulnerabilities found, they enter into the cyber system using some kind of malware.
(2) Access: Usually, the attackers make use of commonly available web apps to communicate with system's command and control server.

(3) Escalation: The hackers continuously observe the network operations, map them and monitor the system with the help of networking and hacking tools and then try to scale the infection.
(4) Exfiltration: By using admin tools, attackers acquire the access to admin's machine or account.
(5) Sustainment: Using remote desktop tools, attackers compromise the infected networked devices and try to access new hosts by penetrating into the internal sub-networks.
(6) Assault: It happens when the hackers alter the functionality of the victim's hardware or disable the hardware totally. This is the rare step during cyber attacks.
(7) Obfuscation: This step makes the victim more confused about the attack by hiding the compromised tracks and also making it difficult to understand how it happened.

2.3 Types of Cyber-attacks

Cyber-attack is an illegitimate attempt to either gain information or monetary benefits. Roughly cyber-attacks are classified as web based attacks and system based attacks. Some people consider three types of cyber-attacks as natural attacks, human blunders or errors, and intentional threats from insiders or outsiders, hackers, and cyber criminals. This section covers various cyber-attacks like backdoors, DoS attack, eavesdropping, spoofing, tampering, repudiation attack, social engineering attack, malware, adware, etc.

2.3.1 Backdoors

Bypassing the conventional entrance to a computing system and creating a new hidden entrance to evade the security policies is called the backdoor attack. In this attack, attacker installs key-logging software or any other software and via which the attackers get access to the victims system. It is a very serious attack because via this attack attacker modifies files, information or installs unwanted software. The famous backdoor attacks include expose of administration and management interfaces, addition of redundant features or functions, creation of hidden parameters, redundant users, authorization for third party access, authentication and authorization between application components, exploit old users in the system to enable identity fraud, flawed hardening, exposure of configuration data, and lack of isolation between different environments (Simsolo 2016).

2.3.2 Denial-of-service Attack

DoS attacks are applied against an availability. In this attack, attacker attacks the system so that actual user can't access the data or resources during this attack. It denies the actual user the use of resources and downs the services of the system during the attack. Generally in a DoS attack, single computing machine or internet connection is used to attack the targeted server by overloading its resources like bandwidth, TCP connection buffers, application/service buffer, CPU cycles, etc. Mobile communication is also greatly suffering due to DoS attacks. Most of the mobile nodes like cell phones; laptops share the physical media they use. Mobile computing resources like bandwidth, CPU, power are constrained. DoS attack can disrupt mobile connection or internet connection by flooding its resources. There are various types of DoS attacks such as direct flooding attack, remote controlled network attack, reflective flooding attack, virus, worm, tear drop attack, protocol violation attack, fragmentation attack, etc. (Neumann 2000). Distributed denial of service (DDoS) attack uses many computing devices and number of internet connections to attack large number of servers by flooding its resources and shutting down them. With the growth in IoT services, the DDoS threat is also increasing. DDoS can work as a subpart of BOTNET by compromising thousands of servers at a time.

2.3.3 Eavesdropping

In this attack, the attacker listens to the system's or network's conversation without their knowledge and uses that conversation for another attacker or enemy of that organization. It is passive kind of attack wherein the eavesdropper just observes and steals the information that computers, smart phones or other network entities are transmitting. Later, it helps in active attacks by providing all the necessary information about the network. There are various programs such as Carnivore and Narus which provide insight that can be used to eavesdrop. Public Wi-Fi networks are the very easy targets for eavesdropping attacks because anyone can join the network by easily available password. Eavesdropping is a serious threat to wireless sensor networks and Internet of Things.

2.3.4 Spoofing

Spoofing is an attack in which the attacker or program acts as if they are the actual, legitimate user of that system or network. They hide their originality from the network and impersonate the system admin or victim. Such threats are frequently initiated via emails where the sender's IP address is spoofed. By stealing and using someone else's identity, attacker monitors network traffic and log information to obtain information about usernames and passwords of the network users. Three types of spoofing attacks are possible including ARP spoofing, IP spoofing, and DNS spoofing. IP spoofing is used in DoS attacks and man in the middle attacks. User never comes to know about spoofing attack because all the packets are received by the intended recipient (Liska 2003).

2.3.5 Tampering

Tampering is a web-based attack in which attacker changes some parameters in the URL of website or path without the user's knowledge. The URL looks legitimate to the user. Hackers perform tampering to gain illegal access to the system or valuable information. With illegal authorization, some parameters entered by the user on particular URL or web page are altered. When some information is being exchanged between client and server, that is manipulated by the hacker. User's privacy is also affected by changing user details.

2.3.6 Repudiation Attack

This is an attack in which a user falsely denies having performed any activity or engaged in any communication. When the user denies for certain actions or transactions, then it is called repudiation. This user may be legitimate user or malicious actor. If the application or system does not have appropriate controls for tracking the user's log actions, then the hacker gets chance to control the information in transit. Robust defense mechanism is necessary which can track and keep record of all user activities. For spoofing mail messages, repudiation attack is used. Repudiation attack greatly affects the performance of mobile adhoc networks (Holkar et al. 2013).

2.3.7 Social Engineering

Social engineering is a type of the nontechnical attack which is based on human mentality. It involves misleading a user and getting useful information from them. Using this information, the attackers bypass the security mechanisms employed in the network. It tempts the user and makes them reveal their secret information thereby breaking normal security process. These attacks include baiting, phishing, spear phishing, pretexting, and scareware. The art of cheating used by malign users to fulfill their greed for money or some other things is nothing but social engineering attack. Phone calls or emails offering the credit cards with unbelievable benefits are the examples of these attacks (FOSSBYTES 2017).

2.3.8 Adware

Adware is a type of software that supports advertisements which are embedded in the application itself. While the program is running, it shows an advertisement. Adware is similar to malware as it uses ads to inflict computers with deadly viruses. Pop up windows continuously appear on the screen where user is working. Usually, malicious adware enters in the system with free software programs and utilities downloaded from Internet.

2.3.9 Ransomware

It is one type of threat, in which the attackers restrict user access to the system and then ask for some amount to remove the restriction. This ransom is paid by online payment methods after which the user can access that service. Systems locked with simple ransomware can be unlocked with the help of security expert. But the advanced ransomware attack encrypts some important files on the victim's system and demands some payment to decrypt and free the files. On 13th May 2017, massive global ransomware "WannaCry" attack moved the world. It struck health care centers, industries, and government offices all over the world by seizing control of affected computing systems until the victims pay a ransom. The ransomware has been stopped spreading with the "kill switch" hidden within the code (Smith-Spark 2017).

2.3.10 Spyware

Spyware is a software which works like a secret agent. The main goal of spies is to gather the information with the use of internet and without knowledge of the user. Spyware may be found in of freeware which are freely available on the Internet to everyone. When victim installs free software, spyware is also installed in the backend via which the attacker gets the information from victim's system.

2.3.11 Scareware

Scareware is a type of threat in which messages suggesting download of useless software pop up on genuine systems. The main aim of Scareware is to create worry among users or victim and to provoke them to download irrelevant software. The popup dialog looks like a system dialog, but it is not same as that.

2.3.12 Phishing

Phishing is a cyber-threat. The main goal of phishing is to get sensitive information like username, password, bank account details, etc. The attacker makes the same copy of an original WebPage and then sends it to the victim. The victim cannot identify the differences because the fake WebPage and the original WebPages both look the same. The victim gives the sensitive information on the matching fake WebPage which are in reality, transferred to attacker's database. The attacker can use this sensitive data for introducing scams. Email phishing is famous nowadays.

2.3.13 Password Attacks

Password attack is performed when an attacker wants to know victim's password. An attacker can guess the password or make a program that can guess the password and tries it on system by bruteforce method, or uses a database which holds a common password and tries it on system or login details of any websites.

2.4 Footprinting

Footprinting is a process of making a map of an organization's network or system. Foot printing is done for information gathering. At first, an attacker has to choose a target system, application or physical location. Once the attacker knows about a target system, he or she can get more specific information from the target.

The main goal of Foot printing is to find details such as the architecture of the network, type of application and server where the important data or information is stored. When an attacker attacks the target, they can know about the operating system, its version, and type of applications, which makes it easy for them to attack the target (Graves 2010).

During footprinting, following type of information is gathered by the attacker.

- Domain name
- Network blocks
- Network services and applications
- System architecture
- Intrusion detection system
- Authentication mechanisms
- Specific IP addresses
- Access control mechanisms
- Phone numbers
- Contact addresses.

Once this information is accumulated, the hacker gets a good idea of the enterprise and the places where the valuable information is kept, and how to gain access to it.

2.4.1 Footprinting Methods

There are several methods for information gathering or Footprinting. Using these following methods anyone can gather the information from the target (Graves 2010).

2.4.1.1 Using Google

Hacker can use Google or Yahoo search engine for gathering the information about the victim. The following are some search patterns that attackers use while surfing for information using Google.

- **Site**: Used for searching a specific website or domain.
- **File type**: For specific file type, a user can mention it here.
- **Link**: For identification of linked pages, it tries to find out among database hyperlinks.
- **Cache**: Web-page version can be identified with it.
- **Intitle**: A specific term in the title of a document is found out.
- **Inurl**: Searches only within the URL of a document.

2.4.1.2 Whois and DNSlookup

Whois

The "Whois" command is used by domains and IP address on the Internet. It is used by system administrators to obtain contact information for IP address assignments or domain name administrators. It works on Linux based systems (Gite 2006).

NSlookup

The user can use the Name Server Lookup (NSLOOKUP) command to find domain names and IP addresses from DNS. If you enter an IP address, its gives you the Domain name. The process is called as a Reverse DNS Lookup. If the user enters a domain name, the user will get back the IP address to which it corresponds; it is called as a Forward DNS Lookup. In practice, NSLOOKUP reaches out over the

Internet to do a DNS lookup from an authorized name server, and then formats the information returned for appropriate display. This command works on Windows as well as in the Linux.

For example:
nslookup google.com

2.5 Wiretapping

Wiretapping is done by placing a monitoring device through a built-in mechanism in other communication technology. Officially, only authorized personnel can tap into live conversations to monitor or even record them. Packet Sniffers, programs used to capture data transmitted on a network, are commonly-used wiretapping tools. Various other tools, such as Wiretap Trojans, are used for different applications.

2.6 Social Engineering

Social engineering is one type of nontechnical attack which exploits human mentality. It is a procedure of misleading the person and getting the information from the users which is useful for hacker or attacker. It is a practice of attackers to deal with the human mind and extract information from the victim, who is useful in bypassing the security mechanism. Social engineering is based on human nature like, being helpful to others, trusting people or fear of getting into trouble (Graves 2010).

2.6.1 Types of Social Engineering

There are main two types of social engineering which are discussed in the following section.

2.6.1.1 Human-Based Social Engineering

- Impersonating an Employee or Valid User
- Posing as an Important User
- Using a Third Person
- Calling Technical Support
- Shoulder Surfing
- Dumpster Diving.

2.6.1.2 Computer-Based Social Engineering

Computer-based social-engineering attacks include the following:

- Email attachments
- Fake websites
- Pop-up windows
- Insider Attacks
- Identity Theft
- Phishing Attacks
- Online Scams
- URL Obfuscation.

2.7 Packet Sniffing

Packet sniffing is known as network monitoring or network analyzing. It can also be used by a network administrator to monitor and troubleshoot network traffic. The network administrator can identify the errors using captured data that can help in maintaining better network transmission or communication (Bradley 2013). In simple words packet sniffer captures all data packets in a given network. Once the raw packet is captured, it is analyzed by the attacker. Via the captured packets, one can know about password or authentication tokens if they are passed in clear text. Hackers can also capture packets for later use. Playback in the replay, man-in-the-middle, and packet injection attacks are the attacks to which some systems may be exposed. Wireshark is one of the most used-tools for Packet sniffing.

Countermeasures for Packet Sniffing

The user can secure himself from packet sniffing by using Encryption such as SSL (Secure Socket Layer) or TLS (Transport Layer Security). This encryption cannot protect from packet capturing, but if the user has used that encryption, then their payload or data is encrypted. So if an attacker gets the packets and finds the payload, then he/she cannot modify that payload because it is encrypted.

2.8 Well Known Ports

Port is an endpoint of a logical connection in TCP/IP and UDP networks. In another words, it can be said that a port is an endpoint of connection between client and server programs on computers in the network. Ports are identified by a Port number which is assigned by the Internet Corporation for Assigned Names and Numbers (IANA) (Webopedia Staff 2015). Some useful port numbers and their functionalities are listed in Table 2.2.

Table 2.2 Port numbers with necessary information

Port number	Port name	Description
20	FTP Data	FTP data port
21	FTP	File Transfer Protocol (FTP)
22	SSH	Secure Shell (SSH) service
23	TELNET	The Telnet Service
25	SMTP	Simple Mail Transfer Protocol (SMTP)
53	DOMAIN	Domain Name Services
80	HTTP	Hyper Text Transfer Protocol (HTTP)
110	POP3	Post Office Protocol version 3
115	SFTP	Secure File Transfer Protocol (SFTP) services
161	SNMP	Simple Network Management Protocol (SNMP)
389	LDAP	Lightweight Directory Access Protocol (LDAP)
443	HTTPS	Secure Hypertext Transfer Protocol (HTTP)

2.8.1 Port Scanning

Port scanning is the process of systematically scanning ports of a computer. Port is an endpoint of any computer from which the information goes in and comes out. Port scanning recognizes open doors or ports to a computer. In network management, Port scanning is legally used but at the same time, some eavesdroppers can use it for the malign purpose.

Types of port scans:

- **Vanilla**—For connection of all the port (e.g., 65,536).
- **Strobe**—For connection to specific ports.
- **Stealth scan**—Mechanisms of scanning which attempts to prevent the request for connection login.
- **FTP Bounce Scan**—For impersonation of cracker's location from FTP server.
- **Fragmented Packets**—Makes packets suitable to pass through firewalls.
- **UDP**—It is the type of scanning for opening UDP Ports.
- **Sweep**—It consists of scanning the same ports of several computing machines.

2.9 Password Vulnerabilities

Passwords are the key information which is needed to access a system. Users select passwords that are easy to guess, like mobile number, birthdate, pet name, etc. Because of this human factor, most of the times, password guessing is successful if some information is identified or easily guessable about the target. Information gathering or Footprinting can be used by a hacker to guess a user's password (Graves 2010).

Types of Passwords:

For password formation, various techniques are available which make use of letters, alphabets, special characters, a combination of letters and numbers, special characters and letters, or the combination of these things. A strong password is hard to guess for the attacker. The following are the rules for creating strong password:

- The password must not contain any part of the user's account name
- It must have a minimum number of eight characters.

It should contain characters from at least three of the following categories:

- Non-alphanumeric symbols
- Numbers
- Uppercase letters
- Lowercase letters.

Types of password attacks:

- **Passive Online attacks**: Eavesdropping on network password exchanges. Passive online attacks include sniffing, man-in-the-middle, and replay attacks.
- **Active Online attacks**: Guessing the Administrator password. Active online attacks include automated password guessing.
- **Offline attacks**: Dictionary, hybrid, and brute-force attacks.
- **Non-electronic attacks**: Shoulder surfing, keyboard sniffing, and social engineering.

2.10 Track Covering

It is the step of the hacking cycle. After the Successful attack or gaining the access to administrator rights on a system, it is necessary to remove the identity or try to cover the track to prevent detection of their presence on the system. To prevent detection, an attacker usually deletes or erases any error messages or security events that may have been logged in. While disabling, auditing and clearing the event logs are a major concerns to cover the track and avoid detection.

Whenever the attacker gains the access to administrator privileges, he first disables the auditing records. Windows auditing records contain events in a log file that are stored in the Window Event Viewer. It includes logging into the system, application, or an event log. An administrator can choose the level of logging implemented on a system. An attacker checks the level of logging and indication of their presence in the system.

It is simple for an eavesdropper to break security logs which are present in the Window Event Viewer. If the event log contains very few entries, then the user should understand that an attacker has cleaned other events. With Auditpol tool, it can be recognized that the auditing is disabled (Rouse 2005). Examples of some track covering tools include Windows NT Resource Kit, elsave.exe utility, Win Zapper, Evidence Eliminator, etc.

2.11 Malware

It is malicious software designed to introduce malign actions in the intended system. There are various types of Malware such as Viruses, Worms, Trojan Horses, etc. Malware can cause huge damage to hard drives. It can delete some documents, or it can collect some sensitive data without any permission. Malware refers to malicious software that is designed to perform unwanted actions on the system. Malware is of many types such as viruses, worms, Trojan horses, etc., which can cause destruction on a computer's hard drive. This malware usually deletes important documents and files from your computing machine. Also, it can passively collect sensitive data from a device without the victim's knowledge. Fedex has confirmed that its services are suffering from malware attacks (ABC News 2017).

2.12 Viruses and Worms

They are used to infect or modify system contents so that eavesdropper can get easy access to the computing system. These viruses and worms act as carriers for Trojans and Backdoors. They are responsible for spreading the infection on a large scale. Worms haves self-replication capability, and they can quickly spread from system to system. Viruses are categorized according to their infection mechanism, as depicted in Fig. 2.1.

2.13 Logic Bombs

Logic bombs are also called as slag code or programming code; added to the software of an application or operating system that lies inactive until an event occurs, triggering the code into action such as terminating the programmer's employment. The malicious software triggered by response to certain event when particular date or time is reached. Logic Bomb waits till you login to banking website or social networking site. This triggers the key logger and sends the victim's credentials to the remote attacker.

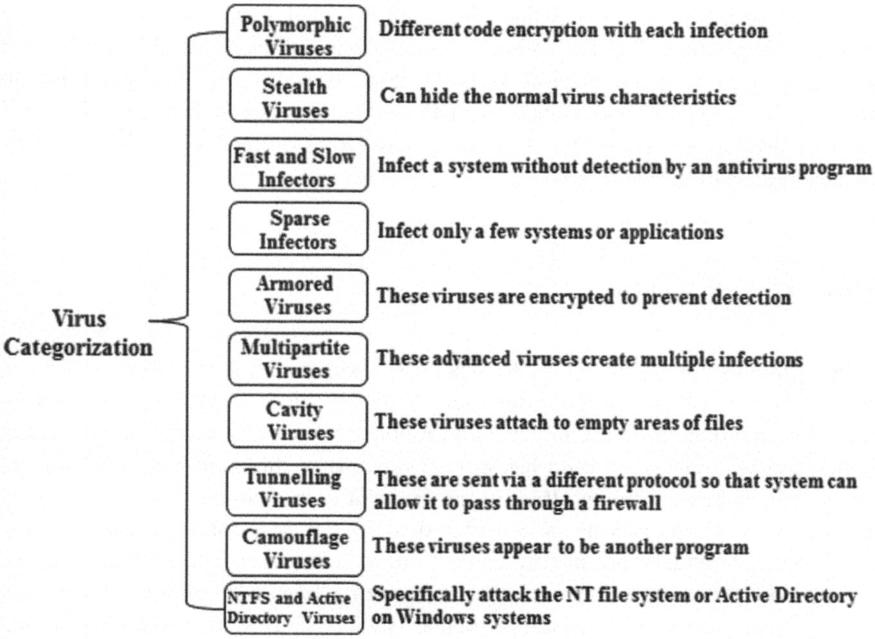

Fig. 2.1 Virus categorization

2.14 BOT and BOTNET

Bot is nothing but a robot or machine with embedded autonomous software codes which make them behave intelligently. BOTs are used by eavesdroppers for automatically posting spam messages on the WebPages opened by the users. Remote attacks are possible with the help of BOTs. Autonomously installing BOTs are more hazardous for computing machines.

BOTNET is a group of infected robot machine systems. BOTNETs are mainly used for introducing DDoS attacks, for producing spam in large amount, for creating Internet marketing scams, for stealing application serial numbers, login IDs, and stealing sensitive information such as credit card numbers, etc. With the BOTNET, within a fraction of time, large computing machine network can be infected or brought to a standstill (Data Security Council of India 2011).

2.15 Trojan Horse

It is a malicious program that masquerades an application and takes a complete control of victim's computer or system. In contrary to the true virus, Trojan horse does not replicate itself by keeping the victim unaware of the attack. These kind of malicious codes are hidden inside some normal mail attachment or some free programs like games. The Trojan horse attacks are as follows. Backdoor or Remote Access Trojans (RATs), Exploit kits, Rootkits, Ransomware Trojans, Banking Trojans, Trojan DDoS and DoS, Trojan Downloaders, Trojan Droppers, FakeAV Trojans, IM Trojans, Game Thief Trojans, Sms Trojans, Spy Trojans, Mail finding Trojans, etc. (Dimitrova 2015).

2.16 Cryptojacking

Cybercriminals use cryptojacking to make easy money with fewer efforts. They use cryptocurrency which is digital currency. With few lines of malign code, somebody's computing machine, tablet, mobile phone or interconnected smart system device can be hijacked. Victim has to bear the cost of the computations and electricity utilized to gain money. Bitcoins are popular cryptocurrency used recently by the hijackers. Cryptojacking is considered to be the new form of Ransomware. These types of attacks are mostly carried out in a browser. These attacks can be detected by giving proper attention to the performance of computing device and various regular activities related to it such as if your device starts suddenly showing high processor usage, or if device is giving slow response than normal or if your device gets overheated during normal operations (O'Brien 2018).

2.17 Supply Chain Attack

Risks in the supply chain network system contains many in between threats such as production units, vendor administration, supply chain quality, transportation security, and many other things related to enterprise regular functionality. Less secure entities in the whole supply chain network are targeted with the malign intent of damaging the enterprise. Supply chain attack can negatively influence almost all the segments including government, financial, oil industry, food industry, pharmaceutical industry, aviation industry, etc. Stuxnet worm and ATM malware are some of the known supply chain attacks. Usually supply chain attack begins with the advanced persistent threat. Information shared with suppliers poses the risk for this kind of attack (Ivanov 2019).

2.18 Summary

Awareness about the cyber threats is the first and most important step in achieving security. After that user can be prepared to combat such threats. To face the cyber attack, the system and user need to design and develop strong vulnerability scanning mechanism, email virus or spam filtering system, personal information and password protection, robust firewall services, etc. Security is not a onetime mechanism. It is a continuous process. So awareness and prepared readiness can save the cyber systems from cyber attacks.

References

ABC News (2017) The latest: FedEx confirms it hit by malware attack. The Associated Press, New York, 12 May 2017. http://abcnews.go.com/Technology/wireStory/latest-uks-health-service-hit-ransomware-attack-47372081

Bradley T (2013) Introduction to packet sniffing. Net security

Data Security Council of India (2011) Cyber crime investigation manual

de Ramos blog V (2016) The sequence of a targeted cyber attack, 20 July 2016. http://www.aim.ph/blog/the-sequence-of-a-targeted-cyber-attack/

Dimitrova M (2015) Types of trojan attacks 2015. Network, browser exploits and security essentials. Sensors TechForum, 31 Aug 2015

FOSSBYTES (2017) What is social engineering? What are different types of social engineering attacks? 28 Feb 2017. https://fossbytes.com/what-is-social-engineering-types-techniques/

Gite V (2006) Linux/UNIX command to find out who owns a domain name. Cybercity

Graves K (2010) CEH: certified ethical hacker study guide. Wiley Publishing, Inc., Indianapolis, Indiana

Holkar AM, Holkar NS, Nitnawwre D (2013) Investigative analysis of repudiation attack on MANET with different routing protocols. Int J Emerg Trends Technol Comput Sci (IJETTCS) 2(3)

Identity Week (2017) Identity Week, Cyber Security News and Analysis for the IT Community, The seven steps of a successful cyber attack. https://www.identityweek.com/seven-steps-of-successful-cyber-attack/

Ivanov I (2019) Cyber security and cyber threats: eagle vs 'new wars'? Academia Publishing

Khanse A (2014) Cyber attacks-definition, types, prevention. The Windows Club, Dec 2014. https://www.identityweek.com/seven-steps-of-successful-cyber-attack/

Liska A (2003) Network security: understanding types of attacks. Pearson InformIT, 13 June 2003. http://www.informit.com/articles/article.aspx?p=31964&seqNum=3

Neumann PG (2000) Denial-of-service attacks. Commun ACM 136. (Academic OneFile)

O'Brien D (2018) The A to Z of cyber security. Symantech. https://medium.com/threat-intel/the-a-to-z-of-cyber-security-93150c4f336c

Rouse M (2005) Port scan. Techtarget

Simsolo Y (2016) COMSEC consulting, the art of securing your business. OWSAP Top Ten Backdoors

Smith-Spark L (2017) CNN, Global ransomware attack: 5 things to know, 13 May 2017. http://edition.cnn.com/2017/05/13/world/ransomware-attack-things-to-know/

The Telegraph, Internet Security (2017) Five of the biggest hacking attacks, 13 May 2017. http://www.telegraph.co.uk/technology/internet-security/9942462/Five-of-the-biggest-hacking-attacks.html

Webopedia Staff (2015) Well-known TCP port numbers. Webopedia

Chapter 3
Phishing

Email or malicious websites are used to collect user login credentials and are provided to the attackers in the phishing attack which is also considered to be the type of social engineering attack. Mid January 2017, many Gmail users all over the world experienced Gmail Phishing attack. The attackers used extraordinarily intelligent mechanism containing a duplicate but looking very real Gmail sign in page. The past email messages and attachments were analyzed to prepare for this phishing attack to send convincing emails to the victim users. Users need to gain awareness about finding real and fake emails by careful observation of the URL (Sulleyman 2017). Here this attack news reveals that phishing is the technique of sending fake emails to the user by spoofing the identity of an established legitimate enterprise or organization with the intent of gathering private and sensitive information from them for malign purposes or gaining money.

Internal Revenue Service (IRS) of America has observed approximately 400% increase in the phishing scams in 2015–2016. Since October 2013, there have been 896,000 phone scam reports, and 5000 victims have paid a combined total of more than $26.5 million due to IRS phishing scams. The users are attracted towards phishing websites due to the lucrative offers provided by the attackers (Addady 2016).

3.1 Introduction

Forged email messages, websites, and phone calls are designed and sent to the users to trap them into giving information about their credit card details or log in details. Main intention behind phishing attack is monetary gain in large amount. Attackers can also install malicious software on your device to steal personal details from your computing machine. Phishing attack comes in various types such as social engineering, link manipulation, spear phishing, clone phishing, voice phishing, etc. (Chaudhry et al. 2016).

R. Prasad and V. Rohokale, *Cyber Security: The Lifeline of Information and Communication Technology*, Springer Series in Wireless Technology,
https://doi.org/10.1007/978-3-030-31703-4_3

At present more commonly seen phishing attacks include deceptive phishing, spear phishing, CEO fraud, pharming, dropbox phishing, and Google Docs phishing. In deceptive phishing, email messages saying to come from genuine sources ask the user to verify their account, re-enter credential information or asking to pay some amount. Users feel it is from legitimate source and are tempted to provide sensitive information. Example PayPal Scammers have sent an attack email asking users to click on the link to resolve some issues related to their accounts. The link was redirecting them to the bogus PayPal login page which was actually collecting user's login credentials and was making it available to the attackers. Spear phishing is a more sophisticated form of phishing. Here the attacker uses more information about the user while writing an email such as user's name, position, organization, company phone number so that the target should feel more connected with the attacker. Users are attracted to click on a fraudulent link or email attachment for getting their personal sensitive data. Spear phishing is a commonly taking place on social media sites like LinkedIn, Facebook, Twitter, Instagram, WhatsApp, etc. (Bisson 2016).

In CEO fraud, the invader poses as a company authority by using email address same as that of CEO or Manager of particular company and writes an email to the targets with the fraudulent email address asking for sensitive data and/or money for some cause. Pharming attackers normally hijack a website's domain name and exploit it to redirect users to a malicious website with the malign intent of catching and stealing the online payments or login credentials to the banking accounts. Google Docs phishing scammer invites target users to view documents on GoogleDocs. The page looks similar to Google Drive but makes the user to enter the credentials and send that information to the attackers. By this, they get access to user's Google account including Gmail, Google Play, Google Plus, etc. (Lord 2018).

Now a day's, detection and identification of any phishing websites in real-time is really a dynamic and complex problem involving many factors and criteria. Because of the uncertainty involved in the detection, fuzzy logic techniques can be effective tools in identifying and assessing phishing websites. Fuzzy logic ensures quality factors rather than exact values. The approach to overcome the fuzziness in the phishing website assessment and an effective and intelligent model for detecting phishing websites is an urgent need of time. This method of phishing detection is based on fuzzy logic and data mining techniques to characterize different phishing website factors. The most important thing is dealing with phishing attack before it actually happens. Phishing attacks come in all shapes and sizes. Victims can be both organizations as well as single users. The organization of the chapter is as follows. Section 3.2 gives information about phishing website. Next Sect. 3.3 discusses about phishing attack technique and how it takes place. Background of phishing websites in explained in Sect. 3.4. Section 3.5 elaborates various phishing techniques. State of the art detection techniques are described in Sect. 3.6. Section 3.7 gives some insights for attack prevention. Last Sect. 3.8 summarizes the chapter.

3.2 Phishing Website

The Internet has become an integral part of almost all our everyday activities including shopping, playing, seeing movies, enjoying music or business activities. But unfortunately, poor security provides a strong motivation for attackers to large financial gains. Emails are not safe because of fraudulent activities on the Internet. Whatever the data is transmitted, sensitive and valuable, it is important to protect confidential information from unauthorized parties or to avoid information from being manipulated. Phishing sites may ask for information such as usernames and passwords, social security numbers, bank account details, personal identification number (PIN), your mother's maiden name, your birthday etc. through emails, advertisements or by mimics of the sites which you usually browse (Basnet et al. 2008).

Phishing websites are the fake websites that are created by attackers, to impersonate genuine websites. Most of these web pages have high visual similarities to scam their victims. Few of these web pages look exactly like the original. Phishing is the exploitation mechanism designed to influence a target to provide personal information and social engineering. Phishing attacks are more recurrent and refined. The impact of phishing is huge because it involves the risk of identity theft and financial losses. Phishing attacks have become a serious problem for online banking and e-commerce users (Egan 2019; Chen and Guo 2006; Rosiello et al. 2007). Figure 3.1 gives the statistics about the phishing attacks during 2018 which is for sure the fact to worry about.

Phishing attacks are normally well-organized and financially motivated crime which attacks target's private information and validation credentials. They not only cause significant loss in the form of money to users, industries and financial organizations, but also damage user's confidence. In case of online banking phishing websites, the targets are tempted to interpret their bank account number, password, credit card number, or other sensitive information. As a result, the user faces compromise of confidential data, and the victims may finally suffer financial loss. Phishing is a relatively new Internet crime in comparison with virus and hacking. Variety of phishing techniques are emerging in this decade (Egan 2019; Dong et al. 2008).

Fig. 3.1 Increase in phishing attack frequency during 2018 (Egan 2019)

Fig. 3.2 Phishing concept

The word phishing comes from the phrase "website phishing", which is a variation of the word "fishing". The idea is that bait is thrown out with the hope just like the fish net that a user will grab it and bite into as shown in Fig. 3.2. In most cases, the attraction is either an instant messaging site or an email, which will take the user to phishing websites (Egan 2019).

Phishing is the act which maliciously tries to gain access to personal information such as usernames, passwords and bank credentials. An attacker exploits electronic communications which are untrustworthy claiming to be from popular legal companies to include social websites, auction sites, online payment processors or IT administrators. Phishing attacks are usually initiated by social engineering and technical scam to steal data related to consumer's personal identity and bank account credentials.

3.3 Phishing Function

Phishing attack is more sophisticated in function in which the fake webpage or email is created that are very similar to the authentic email address and webpage of an industry. The email sent by the attackers normally contains a link which appears to be an official website, which is actually a fake website operated by the fraudster. Once the victim clicks on this website and visits it, any information they enter on that page will be collected by the phisher and may be used for whatever purpose the attacker has in mind. From the beginning to end, the process involves:

Planning—The attack is well planned keeping in mind which company to target by sending spam emails to mass target.

Setup—After appropriate planning, the setup is created including email addresses and websites to target.
Attack—The attacker sends an email or a link through email which poses to be genuine one.
Collection—The malicious actor collects the user information entered into web pages.
Identity Theft and Fraud—The information gathered is used for illegal purposes of gaining money or getting illegitimate identity (Dhanalakshmi et al. 2011).

3.4 Motivation for Phishing

The motivation or the aim of the phisher is normally monetary gains. Similar to monetary gain, few more like identity theft, malware distribution, industrial espionage etc., are the main intents behind phishing attacks (Aggarwal et al. 2013; White et al. 2012). Phishing website is a very big problem, because of its massive impact on the banking and trading segments and because of that, avoiding such attacks is very vital thing nowadays. Various researchers have studied phishing techniques, their characteristics, and have come up with lot of detection mechanisms for phishing attacks (Egan 2019).

Different authors have surveyed available detection methods and anti-phishing solutions. A very simple technique is to prevent phishing at the e-mail level, because most phishing attacks use spam (i.e. broadcast e-mail) to tempt target users to a phishing website. Anti-phishing filters can fight with phishing website at the email level, as it is the primary channel for people behind the phishing websites to reach up to the user. Security toolbar blocks the user activity if the website is identified to be a phishing site. Another approach is to discriminate the phishing sites from the genuine sites. Next technique is two-factor authentication in which the user knows a secret as well as a security token. Two-factor authentication mechanism is a server-side solution, but Phishing can still happen at those websites that do not support two-factor authentication. Insightful information like credit card information is not related to a specific site, making it difficult to protect by this approach (Egan 2019; Mukaram 2014).

According to many researchers, the security tool bars do not effectively prevent phishing attacks. Researchers proposed a scheme that uses a cryptographic identity-verification method which requires changes to both servers and clients. Anti-Phishing Working Group provides a solution directory that consists most of the major anti-phishing companies in the world. They have proposed new methods, but it is vital that web page creators need to adhere to certain rules which creating web pages (Jain and Richariya 2011; Aggarwal et al. 2013).

3.5 Phishing Techniques

At present, malicious people behind phishing website use different methods for phishing. So many researchers have implanted a number of systems by considering number of parameters as input for phishing detection. These people used different algorithms, different software, considering different phishing techniques. Table 3.1 shows detailed summary of different phishing techniques.

Table 3.1 Summary for different phishing techniques

References	Input parameters used for detection	System output	Algorithm	Limitation
Chaudhry et al. (2016)	6 different parameters used for analysis (URL length)	Phishing rate (in %)	Fuzzy logic and Data mining (1) WEKA and CBA packages, (2) WEKA's implementation of RIPPER, (3) PART, (4) PRISM, (5) C4.5	Finding the right feature set is difficult problem using suitable DM algorithm
Egan (2019)	(1) Number of times website visited by user (2) Data submitted by user	Phishing score (0.0–1.0)		UBPD cannot handle all types of authentication credential
Dong et al. (2008)	(1) Visible links (2) Invisible links (3) Unmatching URLs	Warning message		The user will have to use this web browser for opening the emails
Mukaram (2014)	Tweeter specific features—(1) Tweet content, (2) Its length, (3) Age of account, (4) Number of tweet, (5) Follower-followee ratio	Threshold level is used	Machine learning	
Jain and Richariya (2011)	Website characteristics and Image (Screenshot of website)	Numerical value in between 0 and 32		
Aggarwal et al. (2013)	(1) WHOIS, (2) URL, (3) Domain and interdomain, (4) IP address	Warning message	MD5	Longer time span need to be used to gather the websites IP and analyze

3.6 Evaluation of State-of-the-Art Detection Techniques

Initially, researchers had used fuzzy logic and data mining algorithms for phishing detection. They considered six different parameters to decide the phishing rate of a website. The same system can be implemented by considering more number of parameters and also by using more number of rules. It is also possible to use layered structure depending upon the different phishing cases. Also these available techniques are only useful for detection of phishing websites. No single method can provide complete immunity from phishing websites. Table 3.2 illustrates the detailed evaluation of the state of the art detection techniques.

3.7 Insights for the Attack Prevention

It is necessary to find out right feature set in the phishing website using suitable classification algorithms. If maximum number of characteristics and phishing website factors are considered as input parameters for fuzzy inference system and prepare a rule base for all these parameters then it may give high accuracy for phishing detection. Figure 3.3 shows the system approach necessary for phishing detection.

Once the system identifies the phishing website, then the second objective is to design a system which provides better security to the user from the different phishing attacks shown in the Fig. 3.4. Following this is the approach to remove the phishing website.

3.8 Summary

The task of phishing website detection is really very interesting with a never ending possibility of algorithmic variations considering a myriad of combination of the factors. The advantage of the fuzzy approach is that it enables processing of variables whose relationships cannot be defined by mathematical relationships. Fuzzy Logic can incorporate expert human judgment to define those variables and their relationships. User awareness and well-designed usable security mechanisms are very much important in restricting phishing attack from affecting their cyber world.

Table 3.2 State-of-the-art techniques for detection of phishing attacks

References	Input parameters used for detection	System output	Algorithm	Proactive prevention	Limitation	Possible contribution
Egan (2019), Chen and Guo (2006)	6 different parameters used for analysis	Phishing rate (in %)	Fuzzy logic and Data mining (1) WEKA and CBA packages, (2) RIPPER, (3) PART, (4) PRISM, (5) C4.5	NA	Finding the right feature set is difficult problem using suitable DM algorithm	It can become more effective, if 2D/3D membership function is designed
Dong et al. (2008), Dhanalakshmi et al. (2011)	(1) Number of times website visited by user (2) Data submitted by user	Phishing score (0.0–1.0)		NA	UBPD cannot handle all types of authentication credential	
Mukaram (2014), Afroz and Greenstadt (2011)	(1) Visible links, (2) Invisible links, (3) Unmatching URLs	Warning message		NA	The user will have to use this web browser for opening the emails	
Jain and Richariya (2011), Rosiello et al. (2007)	Tweeter specific features—(1) Tweet content, (2) Its length, (3) Age of account, (4) Number of tweet, (5) Follower-followee ratio	Threshold level is used	Machine learning	NA		Neuro-fuzzy approach can be used
Aggarwal et al. (2013), Dhanalakshmi et al. (2011)	Website characteristics and Image (Screenshot of website)	Numerical value in between 0 and 32		NA		Image processing can be used
Aggarwal et al. (2013), Afroz and Greenstadt (2011)	(1) WHOIS, (2) URL, (3) Domain and inter domain, (4) IP address	Warning message	MD5	NA	Longer time span need to be used to gather the websites IP and analyze	

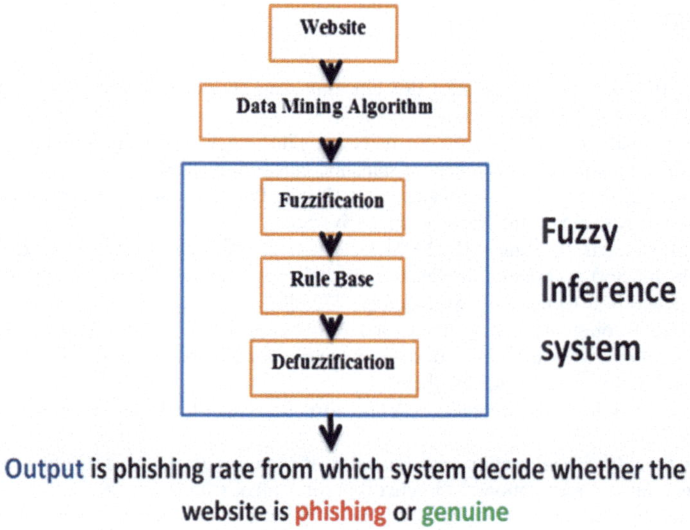

Fig. 3.3 System approach for detection

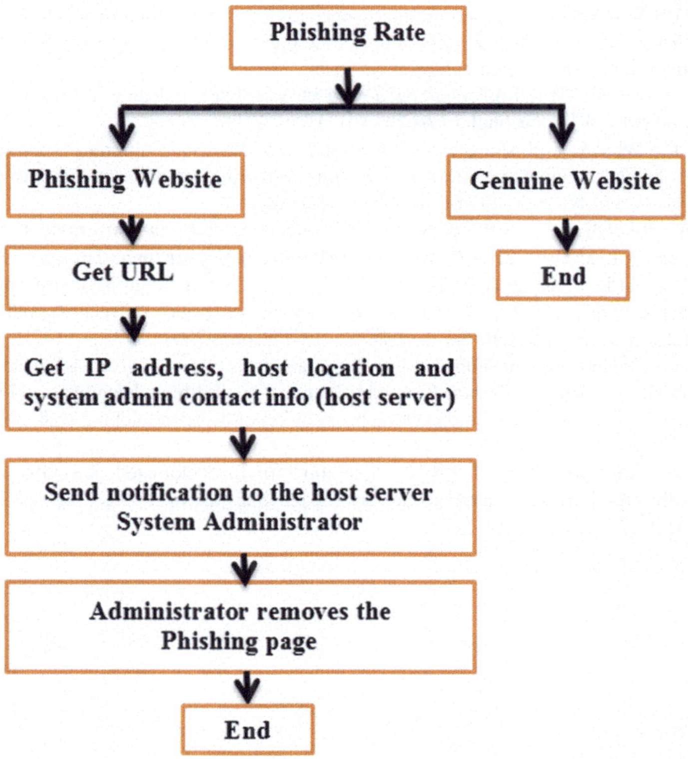

Fig. 3.4 System approach for prevention

References

Addady M (2016) Beware of this latest IRS phishing scam. Fortune 14 Mar 2016. http://fortune.com/2016/03/14/irs-phishing-scam/

Afroz S, Greenstadt R (2011) PhishZoo: detecting phishing websites by looking at them. In: Fifth IEEE international conference on semantic computing (ICSC)

Aggarwal A, Rajadesingan A, Kumaraguru P (2013) PhishAri: automatic realtime phishing detection on twitter. Soc Inf Netw (cs.SI); Phys Soc

Basnet R, Mukkamala S, Sung AH (2008) Detection of phishing attacks: a machine learning approach. In: Soft computing applications in industry, volume 226 of the series Studies in Fuzziness and Soft Computing. Springer, pp 373–383

Bisson D (2016) 6 common phishing attacks and how to protect against them. Tripwire-The State of Security. https://www.tripwire.com/state-of-security/security-awareness/6-common-phishing-attacks-and-how-to-protect-against-them/

Chaudhry JA, Chaudhry SA, Rittenhouse RG (2016) Phishing attacks and defenses. Int J Secur Appl 10(1):247–256

Chen J, Guo C (2006) The detection and prevention of phishing attacks. In: IEEE first international conference on communications and networking in China, ChinaCom'06

Dhanalakshmi R, Prabhu C, Chellapan C (2011) Detection of phishing websites and secure transactions. Int J Commun Netw Secur (IJCNS) I(II)

Dong X, Clark JA, Jacob JL (2008) User behavior based phishing websites detection. In: Proceedings of the international multi-conference on computer science and information technology, pp 783–790

Egan G (2019) State of the phish report: attack rates rise, account compromise soars. Proofpoint, Threat Protection Blog. https://www.proofpoint.com/us/corporate-blog/post/2019-state-phish-report-attack-rates-rise-account-compromise-soars

Jain A, Richariya V (2011) Implementing a web browser with phishing detection techniques. World Comput Sci Inf Technol J (WCSIT) 1(7):289–291

Lord N (2018) What is a phishing attack? Defining and identifying different types of phishing attacks. Digital Guardian Blog. https://digitalguardian.com/blog/what-phishing-attack-defining-and-identifying-different-types-phishing-attacks

Mukaram A (2014) Cyber threat landscape: basic overview and attack methods. Cyber Threat Intelligence, 3 June 2014. https://www.recordedfuture.com/cyber-threat-landscape-basics/

Rosiello APE, Kirda E, Kruegel C, Ferrandi F (2007) A layout-similarity-based approach for detecting phishing pages. In: Third IEEE international conference on security and privacy in communications networks and the workshops, SecureComm2007

Sulleyman A (2017) Gmail phishing: latest cyber attack infects users by mimicking past emails. Independent, 17 Jan 2017. http://www.independent.co.uk/life-style/gadgets-and-tech/news/gmail-phishing-latest-cyber-security-attack-hacking-infect-users-mimicking-past-emails-a7531981.html

White JS, Matthews JN, Stacy JL (2012) A method for the automated detection of phishing websites through both site characteristics and image analysis. In; Proceedings of SPIE 8408, cyber sensing

Chapter 4
BOTNET

In rivalry competition to Mirai Botnet, the second last week of December 2016 experienced a massive 650 Gbps DDoS attack by IoT Botnet named as Leet IoT Botnet. These attacks used large payloads to jam network pipes and thereby bring down the network switches (Seals 2017). Windigo botnet in 2014 infected 10,000 Linux servers and made them send 35 million spam emails per day which affected almost five lakh computers. On the same lines, Grum botnet in 2012 has been found to be responsible for up to 26% of the world's spam email traffic (Thomas 2015).

A Botnet is a network of contagious machines called as 'bots'. The network formed with the help of these infected machines is called as botnets. Botnets are mostly used to launch distributive denial of service (DDoS) and click-fraud attacks against target websites. Botnet attack is very famous in cybercriminals because of its fast spreading. For e.g., Game over Zeus (GOZ) botnet is used mainly to confine financial credentials from infected computing machines, and then use those credentials to redirect those to accounts controlled by cybercriminals. Botnets have becomes dais for the infection of the internet such as spam, e-mails, launch denial of services attacks, click fraud, cyber warfare, cyber sabotage etc. Detection of such Botnets and repairing is of great importance for next generation networks with M2M communication. This chapter elaborates the cyber security essentials for BOTNETs.

This chapter is organized as follows. Section 4.1 covers introduction, evolution and applications of Botnets. Botnet components are described in Sect. 4.2. Section 4.3 discusses Botnet lifecycle. Botnet topology is elaborated in Sect. 4.4. Next Sect. 4.5 describes Botnet detection techniques. Various detection approaches are explained in Sect. 4.6. Section 4.7 covers Botnet prevention. Section 4.8 throws light on practical Botnet activities. Last Sect. 4.9 summarizes the chapter.

R. Prasad and V. Rohokale, *Cyber Security: The Lifeline of Information and Communication Technology*, Springer Series in Wireless Technology,
https://doi.org/10.1007/978-3-030-31703-4_4

4.1 Introduction to BOTNETs

Since the beginning of evolution age, the history had shown tremendous revolutions in science and technology. As the invention of the wheel, for example, had proved to be a milestone in the field of transport, likely the invention of the machine caused a revolution in the manufacturing of goods by minimizing human efforts and energy. At present, we are in the midst of information age. The invention of computers had almost changed the working temperament of a human being. Like any part of machinery facilitates the operation for the production of the machine, similarly, equipment facilitates the functioning power of that particular computer operator. As the time required for any normal human being for solving any mathematical problem is raised by hundred times if the same problem is solved by computer (Alzahrani and Ghorbani 2014).

Nowadays no one can do their official work without the use of the internet. Emails, banking transactions, shopping, etc., have brought the need of internet everywhere. As the internet is widely used, protection of information against online transactions has been a crucial aspect as far as national security is a concern. Uncertified users that can easily access the system have been increased in number. These are usually called as hackers. These hackers have many ways to harm or infect the system by various means. Various risk factors may arise when cyber security is improper. There are attacks like spam, phishing, hacking and many more which can try to access the secured data. Thus, the security of a particular organization or one's personal data can be harmed by bots.

4.1.1 Understanding Botnets

Among all the recent threats to cyber security, Botnets are at the top most lists. They are termed in the most severe threats. The term 'bot' is a short form of 'robot' similar to a fictitious character 'Zombie'. As zombie follows the instruction given by his master and attacks another person making him infectious and zombie. This zombie can again infect another person making him contagious too, and goes on building a huge chain of infected person or zombies.

A botnet mainly contains of a bot master, Zombie pc's and Command and Control (C and C) which relays commands to the bots and report to the botmaster. Various botnets emerged till now are create, PBot, ToxBot, MAchbot, PHP loot, among these Egg Drop is the first botnet. The major activity of these botnets is to send spam, e-mail, conduct distributed denial of service (DDoS) attack and compromise insightful information such as personal banking information and credit card numbers. The intention behind these activities is for monetary gain purpose.

Applications of Botnets

Some of the applications of botnets include spam, DDoS, stealing of confidential information, click fraud, cyber sabotage, and cyber warfare.

Spam: The primary use of spam is of sending spam, e-mails. Botnet named "Grum" approximately sends 40 billion malign e-mails, having 600,000 bots in its network.
DDoS: Commercial DDoS payable service can be used by any user.
Theft of Confidential Information: The botnet named "Zeus" is used to steal banking information.
Click Fraud: Automatically infinite numbers of clicks are generated with the help of click fraud.
Cyber Sabotage: The recent attack using a botnet named Stuxnet came under the act of cyber sabotage.
Cyber Warfare: Government is engaged in strengthening the cyber warfare capabilities which include spreading of malware, distributed denial of service, phishing, for both offensive and defensive purposes (Geers 2011).

4.1.2 Botnet Evolution

First time in the history, Morris worm botnet was invented in the University of Oulu, Finland. In 1989, the first ever bot GM bot was evolved. Pretty park botnet, a password stealing Trojan was evolved in 1999. Following this subseven botnet, a remote controlled Trojan was observed in later part of 1999. GT Bot, which is considered to be a global threat, came with new capabilities such as port scanning, flooding, and cloning in 2000. In 2002, SD Bot was experienced via websites using email and chat for support. After that, Agobot came with modular update to the previous one. In 2003, SpyBot or Milkit botnet with spyware capabilities spread via file share applications and emails. Following SpyBot, the backdoor Trojan on Internet relay chat (IRC) named RBot was observed. PolyBot come up in 2004 with polymorphism capabilities. In 2005, My Bot was evolved which is considered to be new version of SpyBot. P2P based bots like SpamThru, Nugache, and Peacomm were invented and applied by the attackers in 2006. 2007 experienced Storm botnet which was greatly scalable attack. Srizbi and Grum attack occupied almost 60% of spam traffic in 2008. Cutwail and Conficker botnet in 2009 affected the system by almost 46.5% of spam traffic. Stuxnet attack in 2010 targeted Siemens industrial control software in the Uranium Enrichment Infrastructure in Iran. Computer worm, Duqu, evolved in 2011 with Ramnit zero access and Metulji attacks. 2012 experienced attacks like Kelihos, Chameleon. Huge impact malware Windigo was invented in 2014. Mirai malware attack come up in 2016 and after that the huge Leet IoT Botnet hit with massive DDoS attacks. Massive Twitter Botnet, Twitterverse has evolved in 2017. The evolution of Botnet attacks are depicted in Fig. 4.1.

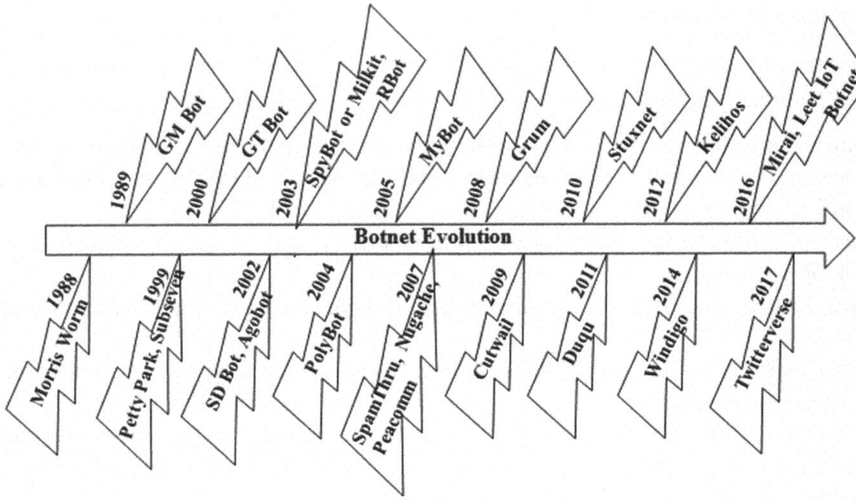

Fig. 4.1 Botnet evolution

4.2 BOTNET Components

Whenever an attack is considered, it is a combination of more than two components. In a similar way, BOTNET is a combination of various things that together do the different work of grabbing the data unknowingly. These components are described in Fig. 4.2.

4.2.1 Bot

The robot is a mechanical device which replaces a human to do a particular task. A robot is a hardware used to work in the physical environment as per the computer program in it. Taking this concept in mind BOT is developed. The combination of two or more BOTs is called as BOTNET. However, BOTs are software material used to carry out a specific task on computers. BOT is a software program made by human to perform a particular task on computer data. This task can be either legal or illegal depending upon what the programmer wants the task to be done. The Internet Relay Control (IRC) channel is very much tedious to control from unwanted material. Therefore legal BOTs are used to avoid unwanted material or user. On the contradictory illegal BOTS are trained to grab the third party information secretively. The communication master who transfers the control information is called as BOTMASTER. It communicates with the BOTs in the network using standard-based network protocols such as IRC and HTTP. The computer on which this BOT is placed and executed can be converted into another BOT.

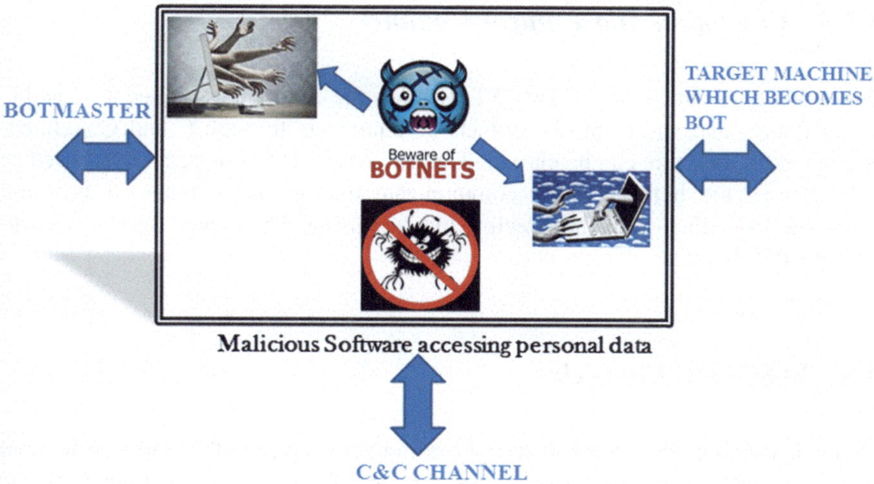

Fig. 4.2 Components of BOTNET

Generally sending a BOT or installing it is done through exploiting web browser vulnerabilities, running a Trojan program which can come from email. Many times the user is unaware of the system being hacked by the BOTs. The BOTs are programmed in such a way that they can destroy themselves after accomplishing the assigned task, or they may also remain in the system to keep them updated and then provide the required information to the BOTMASTER.

4.2.2 Botmaster

As the human is a controller regarding controlling a robot, similarly here the BOTs are controller son behalf of their BOTMASTER. The main machine which is responsible for sending the control signals to the desired system to accomplish the given task is called as BOTMASTER. The first BOTNET was acknowledged and exposed by EarthLink in 2001.

BOTNETs are known for the malware which is used to crate them. However the same type of malware is used by many of the BOTNETs only the entities vary. BOTs are spread through virus and malware. The Hypertext Transfer Protocol (HTTP) network is easily affected by this. Hence, the percentage of the system getting influenced by BOTHEADER (BOTHOST) is more. Many algorithms have been implemented in the detection of BOTNETs. This includes BOTHOUND, BOTHUNTER, Principal Component Analysis (PCA), Fuzzy and Neural Fuzzy systems, machine learning approach, flow-based detection using supervised machine learning (SVM), etc. All these algorithms mainly consist of two important aspects, i.e., training and testing phase. Hence, the accuracy of these algorithms depends upon the quantity of the database used while training.

4.2.3 Command and Control Channel

The server over which the BOTMASTER sends the command is called as C and C server (Command and Control). Botnets communicate through C and C architecture. C and C structure can be classified as IRC based, HTTP-based, DNS-based or P2P. Hence, the botnets which communicate through these types of boat are classified depending on the architecture they are using. The most difficult task is to find out P2P botnets.

4.3 BOTNET Lifecycle

Lifecycle is a journey of the botnet, right from its conception till it achieves the final suspicious task. Figure 4.3 shows the general life cycle of any botnet. While through these stages, if the botnet encounters any failure, it will fail the botnet aim.

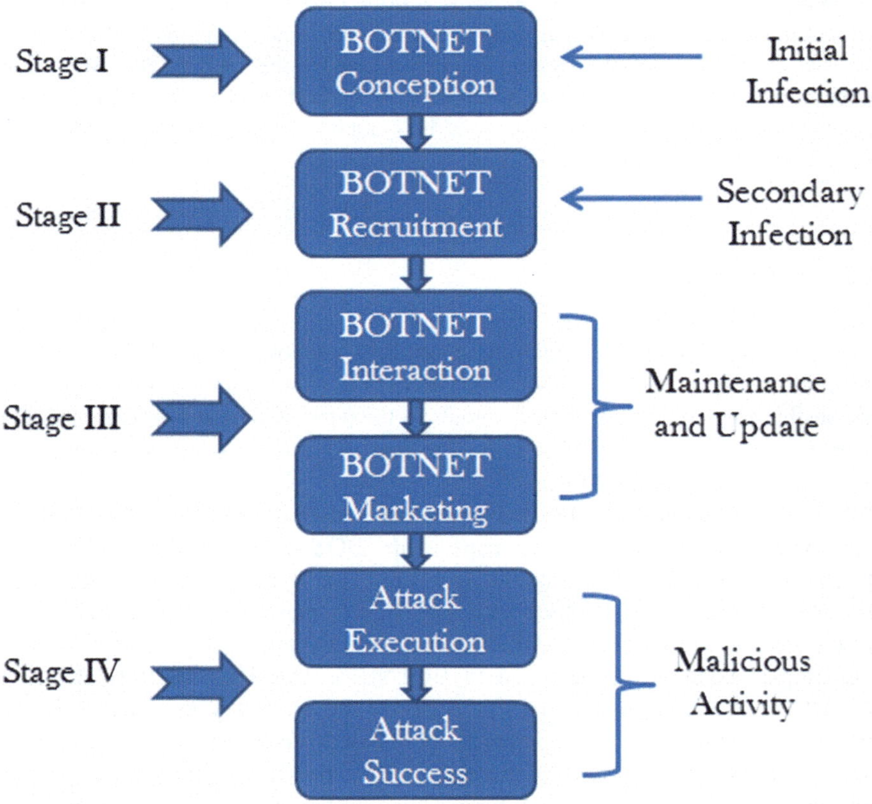

Fig. 4.3 Life cycles of BOTNETs

Stage I i.e. the initial phase consists of a conception of any BOT to achieve the desired purpose. For this, the attacker scans for the system that can be easily infected and can be caught under the master's control. Once this system is found, the master can turn quickly to the victim through various means. In this way the conception of BOT takes place. Now when the bot has been born, it can easily spread infection and also can recruit other infections in it. In this secondary injection stage, the infected machine executes a script, generally called as shell-code.

This shell code fetches the actual harmful material from the bot. This can transfer to locations such as FTP, HTTP or P2P. This material fetched is then installed on the target machine. As soon as the program is installed the target machine now comes under the complete control of the BOTMASTER and turns into a zombie. This zombie is now capable of running the malicious code. When a zombie is formed, it needs a channel through which it can communicate with the master, in order to have maintenance and update of the latest information about the malicious activity to be done. To fulfill this, bot program launches a command and control channel (C and C) and connects the zombie to the C and C server. Now the zombie becomes a part of the attacker and can easily attack another vulnerable host or can also execute the malicious code. The botmaster uses the C and C channel to send all the updated information on the malicious activity. BOT program received are run by the zombies. Thus, the malicious activity is executed, and Botnet achieves success. Further section will give the brief discussion of botnet lifecycle.

4.3.1 BOTNET Conception, Recruitment, and Interaction

There are three main stages of BOTNET formation such as conception, recruitment, and interaction as described in the following subsections.

Conception In the Botnet life cycle, the conception is at the first stage. This stage is divided into three phases such as motivation, design, and implementation. Motivation is required to create a botnet. Hence, there may be different motivation such as money, entertainment, ego, and status. No matter whatever may be the reason, Motivation includes designing and implementation of the desired Botnet. Designing of Botnet is explained in three different styles such as Centralized, Distributed or Hybrid.

In Centralized design, Bot contact C and C server for receipt of information. The advantage of centralized designing is that less time is required for transmission of information. Whereas in distributed architecture all the bots in the botnet act concurrently as server and clients. However, the time needed is much more for transmission. Finally, Hybrid Botnet is a combination of two designs. Here, the existence of one or more centralized servers is there. If in case, one of the servers gets disconnected, the rest of the botnet still continues to work

Recruitment To implement, the designed Botnet it needs to be recruited. Recruitment is also known as infection. The botnet is recruited in such a way that even new and unreported bugs are explored, and also its capabilities are increased. Designing the website in a very attractive way or also sending e-mails with tempting browser links to click on is another recruitment method.

Interaction In this stage, the interaction takes place on the orders sent by the Botmaster, interchange of messages between bots, communication between botmaster to external servers. However, the botnet interaction stage has become a major concern area for the researchers.

4.3.2 BOTNET Marketing

BOTNET Marketing is a stage, where the Botmaster creates and presents the Botnet with useful functions. While another option is to rent the services, there are various prices for renting the services namely;

DDoS: It costs around \$50–\$100 on daily basis, depending on size of the Botnet and strength of the attack.
E-Mail address: Costing \$20–\$100 to just obtain a list of one million e-mail addresses.
Sending Spam E-Mail: It costs \$150–\$200 to send spam email to one million emails.
Fast Flux Networks: \$1000–\$2000 per month amount is paid for hosting Fast Flux services.
Search Engine Spam: To improve website ranking webmasters use a botnet at a cost of \$300 per month (Rodrıguez Gomez et al. 2013).

4.4 BOTNET Topology

BOTNET topologies are mainly categorized as centralized and decentralized systems which are explained in the following subsections.

4.4.1 Centralized

In a Centralized system, there is only a single C and C server, to which all the Bots report and receive the commands. Hence with minimum C and C servers, it is an easy implementation. However also for the entire botnet, C and C server is a single point of failure. To avoid these problems, techniques such as IP Flux, Domain Flux have been adopted by botnets. The examples of these Centralized systems are botnets using HTTP and IRC as means of C and C Communication (Kim et al. 2010).

- **Star Topology**

Many IRC-based botnets have adopted the star topology. It is one of the simplest methods, in which bots directly communicate with the botmaster. Here the communication speed increases between bot and botmaster. The only disadvantage is of the central point of failure if C and C server fail, entire botnet gets down.

- **Hierarchical**

Here, in this system, one or more proxy layers are inbuilt to conceal the botmaster. If in situation one of the proxies is taken down, the rest of the botnet is working. The Hierarchical structure separates out the proxies as a third portion.

4.4.2 Decentralized

Here in this topology, control and command are provided by a more than single entity, instead by a single one. Multiple C and C servers are present to monitor the bots with regardless of master –slave relationship between them (Ollmann and Damballa Inc 2009).

- **Distributed**

In the presence of multiple servers, though they are distributed at different locations, they communicate directly with each other. Bots located at the similar location are connected to nearest C and C server, which provide fast communication, and also, give benefits such as load distribution availability and resilience.

- **Random**

Here, any Bot is used to send a command to other Bots, hence no such master-slave relationship. As the centralized C and C server is not present, it is hard to locate the botmasters or attack it, as here the alternate route is not present, the shutdown of a single Bot does not affect. In this topology an unpredictable delay takes place. As each Bot maintains its peer list, and if one Bot is captured all other Bots are revealed.

4.4.3 Hybrid

Hybrid structure is a combination of centralized and decentralized topologies. Eg., a botnet using centralized structure between C and C server at the front end proxy bots can put peer to peer as C and C for the Bots under the control of individual proxy Bots.

4.5 Botnet Detection Techniques

Various BOTNET detection techniques are discussed in the following subsections including active and passive detection.

4.5.1 Bot Detection

With or without regarding the bot families, the detection of the bot can be performed. Initially, protection of systems and networks from infections is of prime concern. Otherwise, security researchers are more interested in identifying bot families. The vulnerability of a host or network to botnet infection is indicated by detection of bots. Hence, some remedial approaches are made for recovering of botnets from infections. Figure 4.4 shows the botnet detection flow.

- Active Detection: Actively participating in the operation of detection can be termed as Active Detection. It includes approaches such as Infiltration and C and C server Hijack.
- Infiltration: In this method, a defender controlled machine acts as an actual Bot and examines the C and C server or other peers such as of a peer to peer based

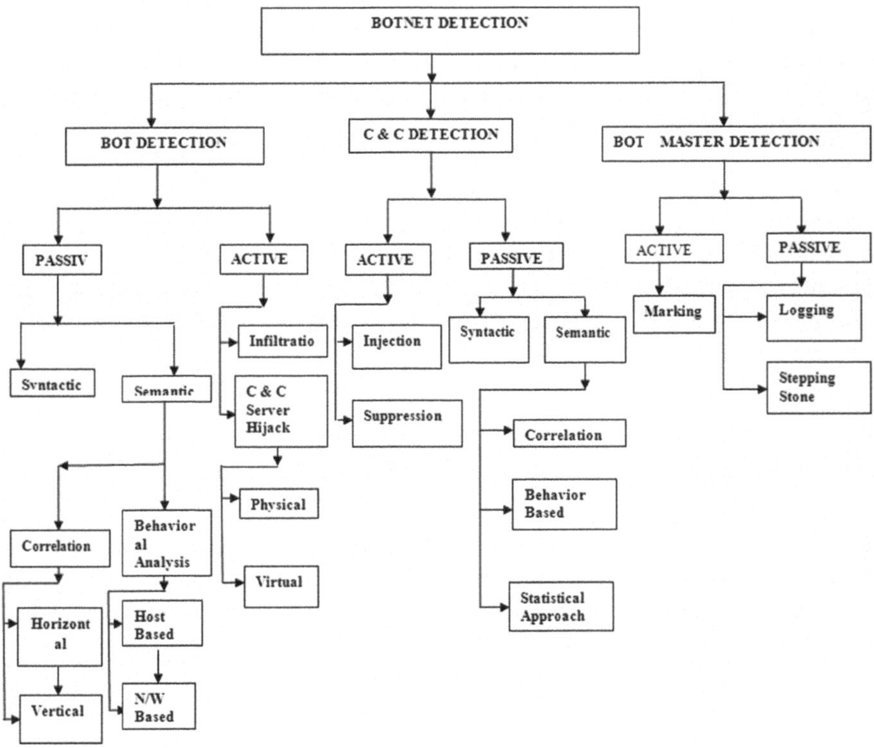

Fig. 4.4 Botnet detection tree (Valeur et al. 2004)

botnet to obtain details of other bots. The defender successively incurs information about other Bots, while disguising itself as a new Bot and also obtains a new peer list.

- C and C server Hijack: The relationship between Bot and C and C server is of a master slave. C and C server acts as a master since bots report to and receive commands from C and C server. If the C and C Server are brought under control, all the information regarding the contacts can be revealed. It is possible only when the rallying mechanisms are exploited. This information can be used to hijack the server. C and C server hijack is possible in centralized topology. Also, physically or virtually the seizure of C and C servers are possible.

If C and C Server are not in the geographical diverse location, during the physical hijack, the law enforcement agencies actually physically take hold of the servers. While, in virtual hijack the directions of C and C Servers are changed and they are bought under the control of defenders. With the help of DNS sink holing data sent by Bots to known botnet can be redirected to defender controlled machine.

- **Passive Detection**: Without actual participation, silently observing and analyzing the activities of botnets are termed as passive Detection. They can be Syntactic and Semantic.
- **Syntactic**: Syntactic is also called as Signature Based Approach. Comparisons of present and pre-determined patterns of botnet infection are done in syntactic approach. The degradation of signature-based detection takes place if strong encryption and bot binary obfuscation take place. Hence, newly born threats go completely undetected if the signatures are not developed.
- **Semantic**: Under this Semantic method use of events and protocols information is observed to detect the malicious behavior. Semantic is further classified as Correlation and Behavioral Analysis.
- **Correlation**: Both Bot and Bot families are used. Hosts, which perform similar activities has been clustered. Bot using both centralized and decentralized topologies, correlation techniques had managed to detect bots all the vital activities such as egg download, propagation activities and C and C communication includes as primary data while attacks and malicious activities are included in secondary data. Correlation can be performed for both, primary as well as secondary. Correlation is again further divided into Vertical and Horizontal Correlation.

Detection of Bots by observing similarities in host behavior and communication is included under horizontal relationship. BotMiner and BotSniffer are the best examples of Horizontal Correlation. In Vertical Correlation, Comparison between single bot and model bot is done. BotHunter and BotTracker are the best examples of the Vertical correlation.

- **Behavioral Analysis**: Analyzing botnet in its machine or traffic behavior from an established pattern is termed as Behavioral analysis. It can be Host-Based or Network Based.

- **Host-based detection**: It examines for a similar sign of bot like behavior as a host. A set of bot activity called as 'Gate functions' are set. If a diverse activity is observed or suspected, then trained data is passed as arguments gate function.
- **Network Based**: Information obtained from network traffic and services under network based methods is used to detect bots.

4.5.2 C and C Detection

C and C channel are an important aspect in Botnet structure and hence its detection. Identification and analysis of C and C assist in understanding botnet behavior. This information helps to identify bots and possibly C and C server.

Active Detection: Active participation in the botnet operation is called positive detection. It involves injection and suppression.

(1) Injection: Injection means providing information also called as packets into different network flow. If an acknowledgment is obtained i.e. reply is obtained for the sent packets, it indicates that the flow is included as a part of C and C communication. Here for C and C detection, reverse engineering is required. Here information can be automatically fed to the detector with information about protocols and known botnets. With the help of this knowledge, information or packets can be injected into the suspicious flow for comparisons with known botnet response. BotProbe is a tool available to identify the same botnet in C and C communication. Bot Probe helps in distinguishing between human and bot responses. Not similar to human, bot response in a deterministic way to typographic mistakes.
(2) Suppression: In Suspicious network flow; the incoming/outgoing packets are suppressed to understand the common responses of the C and C communication. For example, if a suspected bot requests C and C server for the update and its application goes un-responded, but after sometimes bot activates its mechanisms and responses which confirm its infection.

Passive detection: Passive Detection is a complement for active detection. It includes silent observing network traffic, looking for clues of C and C communication. Passive detection is further classified as Syntactic and Semantic.

(1) Syntactic: In this detection, the signature-based model is developed. The frequent recurrence of strings and token, results in the formation of signatures in nasty traffic.
(2) Semantic: Here some heuristic is associated with particular behavior with C and C traffic.
(3) Statistical: It is used to detect botnet C and C Communication. Supervised Learning is prominently used for C and C detection. Features such as range of Packets lengths, Interpacket arrival time, and flow duration are involved.

(4) Correlation Approach: To obtain a pattern in network traffic is the prime idea behind correlation based methods. Strayes and Gu are the best examples of this practice.

Behavioral Based Detection: By observing the difference between the regular traffic for its similarity with existing model, C and C traffic is identified. For example, if a machine is connected to the other port which is oversea, without the user's knowledge, then it is a cause for concern. After being following the C and C protocol, C and C communication can give some solution. Wurzinger is the best example of it.

4.5.3 Botmaster Detection

Very few techniques are there to attack the Botmaster. Botmaster detection creates severe implications on a botnet. Also, a total disbandment for the entire botnet can be caused. Hence, Botmaster is very crucially protected and difficult to detect.

(1) Active Detection: Manipulation is involved in the detection of Botnet activity. Active Botmaster includes Marking Technique.
(2) Marking Technique: It is a technique used to trace the culprits responsible for malicious activity over the Internet. Various available marking techniques include, Probabilistic Packet Marking, the Internet Control Message Protocol (ICMP) Traceback, Deterministic Packet Marking, Ramsbrack is an example for Marking.
(3) Passive Detection: Without manipulation, analysis of data and network traffic involves passive detection of Botmaster.
(4) Logging Detection: In logging, information about packets are recorded by routers. Massive computational complexity and scalability issues are incurred by logging mechanisms. So far logging devices are not used to detect Botmaster.
(5) Stepping Stone Detection: Here, Botmaster hides its identity behind more than one stones. Stepping stone is not the direct method while it is used recursively to identify the Botmaster. The two obstacles in the path of detection are like one is a delay in the arrival of packets to the C and C server while other is the additional packages called 'Chaffs' are added on Botmaster to confuse the detection process. Blum and Zhang are the examples for stepping stones.

4.6 Detection Approaches

For Botnet detection, different solutions have been proposed. Various methods are being categorized as the active and passive analysis. Honeypots and Honeynets are considered under the approach of active analysis while passive analysis includes detection on, Signature based, DNS-Based, Anomaly based and Data Mining based.

4.6.1 Honeypots and Honeynets

The Honeypots is a method where vulnerabilities are initially introduced in the systems, to check or observe the attacks. Honeypots crucially detect security threats, malware signatures and also deeply understand the motivation and technique behind the assault by the bot-header. On a larger area, honeypots of various sizes form honeynets. Linux operating system is more favorable for honeypots than any other system. Depending on emulation capacity, honeypots are segregated into high interaction and low interaction honeypots. High-interaction honeypots simulate all the factors of an operating system. If it is real operating system and also the ports and protocol are known, then high interaction honeypots respond as well as simulate every aspect of it whereas little interaction simulates only essential features of a system. The intruders are permitted to gain control of the system in high interaction whereas little interaction does not allow such outsider.

A system called Network Intrusion Detection System (NIDS) is developed. To examine the traffic on the honeypot, the gathered information is not made available to the system. The data, i.e., signatures of known attacks is required by this newly developed system, NIDS, to detect malicious behavior whereas honeypots detect vulnerabilities which are yet to be found. Here in honeypots, computers are positioned as honeypots. They are placed in the corporate network, in dematerialized zone or also outside a corporate network. Here each area requires a different type of security. Once, the computers are infected by malicious code; they become very harmful for the cooperation. Otherwise, these machines are difficult to reach.

Computers placed in dematerialized zone have some security restrictions whereas those placed in outside zone have fewer security restrictions. For detection and prevention, honeypots and honeynets have become more familiar. Hence, attackers are designing new ways to protect from honeypot traps. By using public internet thread report, the intruders have developed suitable techniques such as VMware or Emulator virtual machine which are successful in finding out intelligent honeypots (Cooke et al. 2005).

4.6.2 Signature Based Detection Techniques

To classify or examine malicious threats, malware executable names are widely used. Executable running on an operating system has a particular power for name based on known malware. Intrusion detection system such as Snort is running for malware signatures. Detection and monitoring signature are done by it. The segregation of signatures is based accordingly to the executable malware or according to the traffic generated by the malicious network. Only known bots can be detected whereas signature-based detection is not possible for unknown bots. On a larger scale, in much intrusion detection system firewall exists. Each system generates thread alerts.

A framework called "BotHunter" (Goebel and Holz 2007) has proposed a system which correlates IDS based detection signals. A network dialogue correlation matrix is used. Here the 'BotHunter' evaluates the IDS discussion into the matrix. After the completion of the event each IDS discussion, weight is calculated by the system. Finally, whether it is a malicious activity or not is decided by the system. A detailed framework for intrusion detection alert correlation is designed. It relies on Sensor Ontology Database. In this Framework Sensor alerts are normalized and preprocessed. Finally, intrusion reports are submitted to the security administrator (Gu et al. 2007).

4.6.3 Anomaly Based Detection Techniques

In the research work of paper (Valeur et al. 2004), algorithm based on statistical techniques for detection of on campus Botnet server has been proposed. BotSniffers also proposed detection techniques in (Binkley and Singh 2006). It is designed to detect botnet command and control channels by using both network behavior anomalies and network channel similarities. Botnet researchers are working on a new platform of botnet detection techniques. Here, researchers have considered methods based on network behavior (Gu et al. 2008).

An algorithm for detection and characterization of Botnets is proposed in (Feily and Shahrestani 2009) based on passive analysis. This approach is based on flow data in the transport layer. As the algorithm does not consider encrypted payload data, this algorithm also detects encrypted botnet interactions. This approach is made simple as compared to others. Simultaneously, the bot reveals network behavior glitch and also communicate with the C and C server.

4.6.4 Data Mining Based Detection Techniques

Anomaly based techniques and network behavior anomalies are growing day by day. Also, C and C data does not reveal strange behavior. So it is very tedious to discriminate between C and C traffic and normal traffic behavior. So to extract the unexpected patterns, pattern recognition, and data mining techniques have powered to be very useful.

Anomaly and data mining based botnet detection system are reviewed and introduced in the research work by (Lu et al. 2011). The passive analysis is a mechanism to detect C and C traffic and is applied in some detection techniques (Davis and Clark 2011). Another static analysis based method is proposed which is based on correlating multiple log files obtained from different network counterparts, wherein, IRC based as well as non-IRC based both systems are applicable (Strayer et al. 2008).

Payload and traffic detection flow can be achieved by a new proposed system based on N-gram feature selection and is described in (Feily and Shahrestani 2009). Based on N-gram features, clustering each request is done to detect anomalous behavior. BotMiner can detect real world botnets including IRC based, HTTP-based and P2P botnets through very less false positive rate (Masud et al. 2008).

Table 4.1 shows the recent research being carried out in the detection of botnets. The important threat for security is the malicious software commonly known as the botnet. Therefore, the detection of this kind of threat is a crucial measure to be taken. Botnets spread through HTTP protocol has been detected by comparing benign HTTP clients (Zarras et al. 2014). This algorithm used here is called as BOTHOUND, and it detects network-level HTTP based malware. Botnets can also occur in mobile. In this case, it is a collection of compromised nodes, and they perform coordinated attacks. Mobile bots do not require centralized nodes; they work with the monitoring of any endangered node around (Garant et al. 2013). These botnets are comparatively slower than Internet botnets. According to the studies they have estimated the increase in the size of botnet over the time. Cloud computing is another giant data storage block where a large amount of reliable information is stored. This room can be used other than authorized users. TCP flood and UDP storm base botnets behavior are monitored and studied by many researchers (Lu et al. 2014). This paper compares the certain workloads using Principal Component Analysis (PCA). Memory consumption and I/O activity are studied in this bimodal distribution. Corporate area network is a computer network made up of an interconnection of local area networks (LAN) which can also be easily affected by the malicious software. Fuzzy and neural fuzzy techniques have been used for the detection of botnets (Badis et al. 2014).

This is achieved with the help of multiple agents involved in communication with the corporate network. In 2007, Collins et al. worked to detect future botnet address with the help of corrupt system (Savenko et al. 2013). Detection technique of botnets is classified as structural based and behavior based. Structural based detection use distinctive characteristics used by cyber criminals, and therefore the results of these techniques are not so good. Signature-based and DNS based detection comes under this category.

Signature-based detection is a useful detection technique for known bots. DNS traffic anomalies can be detected by investigating the DNS traffic(Snort 2006). Behavior Based Detection is further classified as an anomaly based and communication pattern based techniques. Anomaly based detection method is based upon the signals received which reflect the bot characteristic such as high network latency, high volumes of traffic, traffic on unusual ports, and abnormal system behavior in the network.

Some of the companies have implanted rules like anti-spam law, Computer Information System Security Protection Rules and Regulation, EU Privacy and Electronic Communications Directive. These rules help the system to avoid unwanted emails from new BOTs. Companies such as Google, Cisco Systems, Microsoft, and Symantec have contributed to the detection of BOTNETs.

Table 4.1 Comparison of botnet detection techniques

Detection approach	Unknown bot detection	Protocol and structure independent	Encrypted bot detection	Real-time detection	Low false positive
Signature based (Gu et al. 2008; Snort 2006)					
	Yes				
Anomaly-based (Valeur et al. 2004; Binkley and Singh 2006; Ji et al. 2013; Karasaridis et al. 2006)	Yes				
	Yes		Yes		Yes
	Yes		Yes		Yes
	Yes	Yes		Yes	Yes
DNS-based (Narang et al. 2013; Dagon 2005; Kristoff 2004; Schonewille and van Helmond 2006; Choi et al. 2007; Zand et al. 2014; Zargar et al. 2013; Mendonça and Santos 2012; Al Ebri et al. 2013; Lu et al. 2014)	Yes		Yes		
	Yes		Yes		
	Yes		Yes		Yes
	Yes		Yes	Yes	
	Yes	Yes	Yes		Yes
	Yes	Yes			Yes
	Yes	Yes		Yes	
	Yes	Yes		Yes	
	Yes	Yes			
Mining-based (Cooke et al. 2005; Masud et al. 2008; Stevanovic and Pedersen 2014; Strayer et al. 2008; Masud et al. 2008)	Yes				
	Yes				
	Yes	Yes	Yes		Yes
	Yes	Yes	Yes		Yes
	Yes		Yes		

4.6.5 Dedicated Laws for Botnet

The countries such as Italy, Britain, Denmark, Spain and others that are members of EU should not send business email in EU without the previous agreement based on this treaty (EU 2008) recipient. Table 4.2 shows the laws which have been implemented for protection against botnets.

The US government is taking serious steps against the cyber threat related to botnets. Administrative officials belonging to US President's team have declared that the government had started Industry Botnet group (IBG), a coordinated project that involves private enterprises and trade units. Table 4.3 shows different methodologies as well as various detection techniques implemented in various research papers along with result and future scope. The table can thus give a brief idea about the recent studies been carried out on different botnet research areas.

Table 4.2 Rules and regulations implanted by different countries

Rules and regulation	Country	Act
Computer Information System Security Protection Rules and Regulation	People's Republic of China	1994 (APEC 2008)
Anti-spam law (The Law on Regulation of Transmission of Specified Electronic Mail) to address the danger of spam	Japanese Ministry of Internal Affairs and Communication	APEC (2008)
Act on Promotion of Information and Communication Network Utilization and Information Protection	Republic of Korea	2001 (APEC 2008)
EU Privacy and Electronic Communications Directive	European Union	July 2002, (APEC 2008)

Table 4.3 Different detection methodologies and future scope

References	Methodology	Detection/technique	Result/conclusion	Future scope
Zade and Patil (2011)	• Prevent from becoming Zombie, • Detect them, • Take down the botnets	Prevention by, • Honey pots • IRC Tracking • DNS Tracking	Botnets a platform for cyber sabotage and cyber ware	
Wang et al. (2009)	Stability detection algorithm	Analysis on control flow stability and storm worm C and C communication	Both encrypted and unencrypted data identified using proposed approach	
(Pieterse and Olivier 2012; Ramachandran and Dagon 2006)	SMS Botnet detection framework	Signature based detection and anomaly based methods		To test the viability of framework and also efficiency, scalability, and accuracy using large datasets
Derhab et al. (2014)	Security framework named Spam Trapping System(STS)	Prevent–then-detect approach	STS incurs a very low detection time and provide better performance	Extend proposed framework to detect the botmaster and C and C server
Ullah et al. (2013)	Android botnet development model and Android botnet discovery process	Identify features as, repackaged application, receiving commands, messaging, steal information, third party application markets, additional content downloaded	Features help in detection and rise of new android botnets	Focus on identification of android botnets by means of signature based and/or a behavior based detection model

(continued)

Table 4.3 (continued)

References	Methodology	Detection/technique	Result/conclusion	Future scope
Lu and Brooks (2012)	Use of probabilistic context free grammars (PCFGs)	LALR (Look ahead-left to right) parser is used; statistical X^2 test is used	Can be easily extended to other systems	Coherent considerations, to make system more resilient and secure
Karasaridis et al. (2006)	Use of anomaly based classification of intrusions	Correlation based feature selection (CFS), consistency based subset evaluation (CSE), PCA, are machine learning techniques	Almost equivalent accuracy and models built with reduced feature set less time	Use of advanced space efficient data structures like bloom filters
Choi et al. (2007)	New botnet C and C signature extraction approach	Two steps: • Extract all frequent strings in network • Assign a score to each string	Real world applicability and can extract meaning C and C signatures	Use system when C and C is encrypted, removal of irrelevant string
Zand et al. (2014)	Research and tests modeling both normal and abnormal activity of networks. Detection framework prototype is proposed	The developed fingerprinting and corresponding visualization method	ROC plots, related charts, and blacklists used were useful in the evaluation of the system	Analysis of alternate statistical approaches such as random forest tree
Mendonça and Santos (2012)	Trust based model that uses cooperative game theory to cluster trusted hosts	Improve the detection score compared to the traditional correlation model	Higher the reputation, trustworthiness are regarded to a host	
Al Ebri et al. (2013)	Novel flow based detection system relies on supervised machine learning	Eight highly regarded Machine learning algorithms indicating best performing ones	For accuracy, traffic flow to be monitored for only limited time period and no. of packets per flow	Optimization of traffic analysis and deployment of the detection approach in online fashion
Klaper and Hovy (2014)	Creating taxonomy of cyber-security, building a Portal for websites security	Browser plugin, personal cyber security assistant Portal	If Politicians, officials, and users understand the dangers, open Government can be protected	Expansions of taxonomy providing more depth, more versatile platform

4.7 Botnet Prevention

Though some of the laws have been implemented to protect against botnets, prevention is always essential to avoid the personal system being attacked by bots. One of the measures for the same may involve updating to the latest updates of antivirus. Also soft wares which are not in use must be uninstalled. The best way to protect your system is to have string passwords and frequently same passwords for different links must be avoided. Every website has default spam mails. One should not click untrusted sites or links. Periodically checking of sent items is necessary. When not in use, the Internet must be disconnected from your system.

4.8 Practical Botnet Activities

Various criminal activities can be carried out through botnets. These activities may include harming personal data, financial effects or even to create a new bot. Some of these attacks are listed below.

- Botmaster asks the Botnet nodes to send or download the fake data thereby flooding air interface (Lu and Brooks 2012).
- In the work by (Khosroshahy et al. 2013), the authors have provided improvements in a framework that are capable of stopping DDoS attacks before entering into victim premises, ease of tracking and reporting the compromised machines for further cleanings.
- Individual infected zombie computers can be employed for many activities that are risky to the enterprise and its workforce (Sadeghian and Zamani 2014).

Table 4.4 gives in detail the basic activities that a botnet can carry out by a zombie computer with the help of commands given through the master.

Table 4.4 Botnet activities (Zombie computer) (Stawowski 2014)

Source of botnet activities	Activity carried out
Email accounts	Distribute malware to the company's customers and partners
Social networks accounts	Distribute hostile links to friends as well as business partners
Smartphones	Send expensive text messages premium messaging
e-banking account, e-commerce account, or credit card numbers	Steal money
Blackmail	For regaining control of computer or data—ransomware for non-disclosure of private or intimate photo-scam jacking

4.8.1 Defense Mechanisms for Botnets

DDoS attack is one of the major attacks caused by botnets. Several defense techniques for the same have been implemented and are under research. HIF technique maintains the database called as IAD (IP address database) for legitimate IP address which is involved in the network in the last two weeks. Whenever a packet is being lost HIF technique is activated wherein it discards packets whose IP address is missing in the database (IAD). The IAD maintains the database of the IP address which has successfully completed TCP handshake. Another defense mechanism is a kind of killbots. Here when a large amount of crown or data packets accommodate at the server, the server switches to suspected mode. It remains in this mode till the server traffic falls to normal and then it comes back to its normal mode.

4.9 Summary

Botnet crime is on the verge of extensive spreading, and also it causes huge damage to personal property of an individual. From the above survey regarding botnets and their detection approaches, research has been more focused on HTTP and TCP based models. Still there is a lot of scopes for the investigation team to give attention to other models as well as algorithms which would work on multi-networks. A lot of stress should be given to laws that can be implemented against botnet crimes. This can help in preventing the botnet crime from spreading, and legitimate action can provide a warning to the BOTMASTER, this, in turn, may reduce the BOT attacks.

References

Al Ebri N, Otrok H, Mourad A, Al-Hammadi Y (2013) Botnet detection: a cooperative game theoretical correlation-based model. In: Third international conference on communications and information technology (ICCIT)

Alzahrani AJ, Ghorbani AA (2014) SMS mobile botnet detection using a multi-agent system: research in progress. A CySe'14, ACM, France

Badis H, Doyen G, Khatoun R (2014) Understanding Botclouds from a system perspective: a principal component analysis. In: IEEE network operations and management symposium (NOMS)

Binkley JR, Singh S (2006) An algorithm for anomaly-based botnet detection. In: Proceedings of USENIX steps to reducing unwanted traffic on the internet workshop

Choi H, Lee H, Lee H, Kim H (2007) Botnet detection by monitoring group activities in DNS traffic. In: Proceedings of 7th IEEE international conference on computer and information technology

Cooke E, Jahanian F, McPherson D (2005) The zombie roundup: understanding, detecting, and disrupting botnets. In: ACM USENIX workshop on steps to reducing unwanted trace on the internet SRUTI, vol 7, pp 39–44

Dagon D (2005) Botnet detection and response, the network is the infection. In: OARC workshop
Davis JJ, Clark AJ (2011) Data preprocessing for anomaly based network intrusion detection: A review. Comput Secur 30(6–7):353–375
Derhab A, Fahad AB, Khurram BMM, Xiang KY (2014) Spam trapping system: novel security framework to fight against spam botnets. In: IEEE 21st international conference on telecommunications (ICT)
Feily M, Shahrestani A (2009) A survey of botnet and botnet detection. In: NAv6 IMPACT research team Kuala Lumpur, Malaysia
Garant D, Lu W, Keene USNH (2013) Mining botnet behaviors on the large-scale web application community. In: 27th international conference on advanced information networking and applications workshops (WAINA)
Geers K (2011) Strategic cyber security. CCDCOE, NATO Cooperative Cyber Defense Center of Excellence, 19–22 June 2011
Goebel J, Holz T (2007) Rishi: identify bot contaminated hosts by IRC nickname evaluation. In: Proceedings of 1st workshop on hot topics in understanding botnets
Gu G, Porras P, Yegneswaran V, Fong M, Lee W (2007) BotHunter: detecting malware infection through IDS-driven dialog correlation. In: SS'07 proceedings of 16th USENIX security symposium
Gu D, Zhang J, Lee W (2008) Botsniffer: detecting botnet command and control channels in network traffic. In: Proceedings of 15th annual network and distributed system security symposium (NDSS'08)
Gu G, Perdisci R, Zhang J, Lee W (2008) Botminer: clustering analysis of network traffic for protocol- and structure independent botnet detection
Ji Y, He Y, Li Q, Guo D (2013) BotCatch: a behavior and signature correlated bot detection approach. Jilin University
Karasaridis A, Rexroad B, Hoeflin D (2006) Wide-scale botnet detection and characterization. In: Proceedings of 1st workshop on hot topics in understanding botnets
Khosroshahy M, Qiu D, Mehmet Ali MK (2013) Botnets in 4G cellular networks: platforms to launch DDoS attacks against the air interface. In: 2013 international conference on selected topics in mobile and wireless networking (MoWNeT)
Kim W, Jeong O-R, Kim C, So J, Seongnam G-D (2010) On botnets, Korea, ii WAS-2010, Paris, France, pp 461–701
Klaper D, Hovy E (2014) A taxonomy and a knowledge portal for cybersecurity. In: ACM proceedings of the 15th annual international conference on digital government research, pp 79–85
Kristoff J (2004) Botnets. In: 32nd meeting of the North American network operators group
Lu C, Brooks RR (2012) Timing analysis in P2P botnet traffic using probabilistic context-free grammars. In: CSIIRW '12, USA
Lu W, Rammidi G, Ghorbani AA (2011) Clustering botnet communication traffic based on n-gram feature selection. Comput Commun 34:502–514
Lu Z, Wang W, Wang C (2014) How can botnets cause storms? Understanding the evolution and impact of mobile botnets. In: Proceedings of IEEE conference on computer communications (INFOCOM '14)
Masud MM, Gao J, Khan L, Han J (2008) Peer to Peer botnet detection for cybersecurity: a data mining approach. In: CSIIRW 2008
Masud MM, Al-khateeb T, Khan L, Thuraisingham B, Hamlen KW (2008) Flow-based identification of Botnet traffic by mining multiple log file. In: Proceedings international conference on distributed frameworks and applications
Mendonça L, Santos H (2012) Botnets: a heuristic-based detection framework. In: SIN'12, India, pp 25–27
Narang P, Reddy JM, Hota C (2013) Feature selection for detection of Peer-to-Peer botnet traffic. In: COMPUTE'13, Vellore, Tamil Nadu, India, pp 22–24
Ollmann G, Damballa Inc (2009) Botnet Communication Topologies—Understanding the intricacies of botnet Command-and-Control. In: WP Botnet Communication Primer

Pieterse H, Olivier MS (2012) Android botnets on the rise: trends and characteristics. In: IEEE information security for South Africa (ISSA)

Ramachandran NFA, Dagon D (2006) Revealing botnet membership using DNSBL counter-intelligence. In: Proceedings of 2nd workshop on steps to reducing unwanted traffic on the internet

Rodrıguez Gomez RA, Macia Fernandez G, Garcia-Teodoro P (2013) Survey and taxonomy of botnet research through life-cycle. ACM Comput Surv 45(4):10

Sadeghian A, Zamani M (2014) Detecting and preventing DDoS attacks in botnets by the help of self triggered black holes. In: 2014 Asia-Pacific conference on computer aided system engineering (APCASE)

Savenko O, Lysenko S, Kryshchuk A, Klots Y (2013) Botnet detection technique for corporate area network. In: IEEE 7th international conference on intelligent data acquisition and advanced computing systems (IDAACS)

Schonewille A, van Helmond DJ (2006) The domain name service as an IDS. Master's Project, University of Amsterdam, Netherlands

Seals T (2017) Leet IoT botnet bursts on the scene with massive DDoS attack. Infosecurity Magazine News, 3 Jan 2017. https://www.infosecurity-magazine.com/news/leet-iot-botnet-bursts-on-the-scene/

Snort IDS (2006) Snort IDS web page. http://www.snort.org

Stawowski M (2014) Practical defense-in-depth protection against botnets. ISSA Senior Member, Poland Chapter

Stevanovic M, Pedersen JM (2014) An efficient flow-based botnet detection using supervised machine learning. In: International conference on computing, networking and communications (ICNC)

Strayer W, Lapsely D, Walsh R (2008) Botnet detection based on network behavior. In: Botnet detection, Springer, Berlin, pp 1–24

Strayer W, Lapsley D, Walsh B, Livadas C (2008) Botnet detection based on network behavior. In: Advances in information security. Springer, Berlin

Thomas K (2015) Nine bad botnets and the damage they did, we live security blog on security news, views, and insight from the ESET experts, 25 Feb 2015. http://www.welivesecurity.com/2015/02/25/nine-bad-botnets-damage/

Ullah I, Khan N, Aboalsamh HA (2013) Survey on botnet: its architecture, detection, prevention and mitigation. In: 10th IEEE international conference on networking, sensing and control (ICNSC)

Valeur F, Vigna G, Kruegel C, Kemmerer R (2004) Comprehensive approach to intrusion detection alert correlation. IEEE Trans Dependable Secure Comput 1(3):146–169

Wang B, Li Z, Tu H, Ma J (2009) Measuring Peer-to-Peer botnets using control flow stability. In: IEEE international conference on availability, reliability and security

Zade AR, Patil SH (2011) A survey on various defense mechanisms against application layer distributed denial of service attack. Int J Comput Sci Eng (IJCSE) 3(11)

Zand A, Vigna G, Yan X, Kruegel C (2014) Extracting probable command and control signatures for detecting botnets. In: SAC'14, Gyeongju, Korea, pp 24–28

Zargar ST, Joshi J, Tipper D (2013) A survey of defense mechanisms against distributed denial of service (DDoS) flooding attacks. In: IEEE communications surveys and tutorials

Zarras A, Papadogiannakis A, Gawlik R, Holz T (2014) Automated Generation of models for fast and precise detection of HTTP-based malware. In: Twelfth annual international conference on privacy, security and trust (PST)

Chapter 5
Malware

FedEx Corporation, a global courier system, Russian criminal investigation agency and a top mobile operator, Megaphone, United Kingdom healthcare centers, worldwide banking services have come under recent malware attack during massive WannaCry Ransomware attack wave (Associated Press, ABC News 2017). Nowadays our society is successfully going towards tech savvy mode. This is very positive step towards growth, but at the same time our infrastructure relies on technology as well as computers. A threat to the computing system has become a threat to the society. There are four key threats to consider like Spam, Bugs, Denials of service, malicious software, etc.

This chapter aims to study malware and the techniques which can be used to detect, confine and wipe out it. Malware has the connection with remaining threats listed above. Malware is normally propagated using spam email, and is also used to send spam. It takes advantage of bugs. Malware may be used to build up denial of service attacks. Malware has become very crucial part in the cyber security. For example, computers connected via LAN or WAN begin experiencing problems as a result of malware that have been downloaded from the host computer. These problems can be very small that they create difficulty in accessing some file or very large that they can damage a whole system.

5.1 Introduction

Hackers exploit vulnerabilities for injecting malicious software into existing website software resulting in identity theft, financial ruin, data stealing, etc. Malware is malicious software which is used to interrupt any computer system operation and gather sensitive operations and can gain access to other systems, display unusual and unwanted posts and advertising. Malware gets installed on user's machine and performs unusual tasks may be of some outsider's benefits by getting the confidential data. Some malwares are also designed in such a way that they transmit the

R. Prasad and V. Rohokale, *Cyber Security: The Lifeline of Information and Communication Technology*, Springer Series in Wireless Technology,
https://doi.org/10.1007/978-3-030-31703-4_5

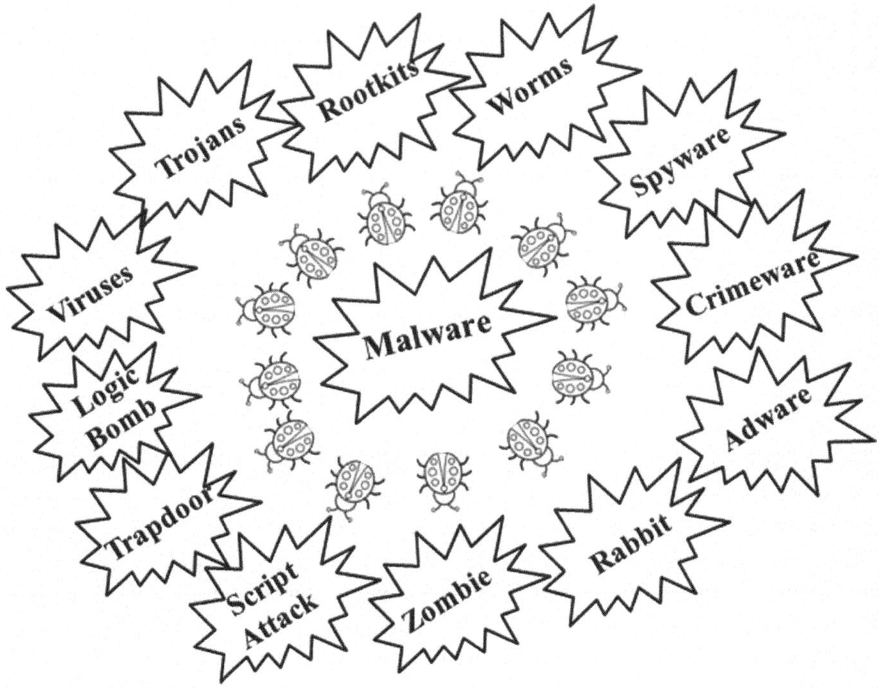

Fig. 5.1 Types of malware

information constantly about your web browsing activity to another party which may be of their known interest.

Malware refers to various intrusive softwares which contain computer viruses. There are various types of malwares like trojan, worms, trapdoor, spyware, adware, shareware, rootkit, viruses, logic bomb, script attack, zombie, crimeware, rabbit, etc. as shown in Fig. 5.1. These programs on running can, in fact, turn to the executable program and scripts which run in the background process unaware of the user. The malware is installed in the software by the intruder. When user installs particular software, the malware gets executed and jumps in the user computing system and it starts transmitting the information or logs to the particular system.

This chapter is organized as follows. Section 5.2 discusses the evolution of malware. Different malicious softwares are described in Sect. 5.3. Section 5.4 elaborates malware investigation mechanism. Deception methods are explained in Sect. 5.5. Section 5.6 throws light on malware detection and analysis. Details about use of virtualization to eliminate malware are given in Sect. 5.7. Last Sect. 5.8 summarizes the chapter.

5.2 Malware Evolution

Elik cloner was the first malware observed in 1981 which infected Apple IIC machine's operating system, Apple DOS 3.3. This malware was attached to game and infected Apple's boot sector and had capability to clone itself to new disks in contact with the infected system. Year 1986 witnessed Brain virus which also infected boot sectors in contact computing systems. Morris worm evolved around 1988 and it brought buffer overflow technique which can replace the machine's code into malicious software thereby executing itself in the new memory space.

Disk killer malware came in 1989. In 1990, polymorphic virus capable of rewrite themselves struck the computing world. Michelangelo virus was observed in 1991 that infected the DOS system by corrupting the boot sectors of the storage devices. Microsoft word and excel were affected by the malware that run inside embedded executables and macros shocked the world in 1995. Botnet concept emerged in 1995 with the appearance of Melissa virus. Trojan came around 1997. Moonlight Maze malware struck the world with nation state cyber espionage by stealing important military technology documents. Social engineering attacks started in 2000 with famous and dangerous love letter worm. Also, Pikachu virus was developed aiming for attacking children gadgets. Anna Kournikova worm came in 2001 with malicious contents hidden inside celebrity photographs and information database. DOS attack known as SQL Slammer evolved in 2003 thereby putting down the Internet all over the world. SpyBot, Milkit malware have spying capabilities which get spread through file share applications. Rbot is the backdoor Trojan on Internet relay chat (IRC).

Similar to AgoBot, the PolyBot was evolved with polymorphism capabilities in 2004. New version of SpyBot came in 2005 which uses hybrid coding and is spread via file sharing applications and e-mail. 2006 witnessed P2P Based Bot like SpamThru, Nugache, and Peacomm. Storm Botnet emerged in 2007 with high scalability and increased resilience. Grum malware which occupied 18% spam email traffic was observed in 2008. It followed Srizbi, Kraken, and Maripose. 46.5% of the spam traffic was generated by Cutwail malware in 2009. The massive cyber physical attack Stuxnet took place in 2010 targeting Uranium Enrichment infrastructure in Iran. Computer worms like Duqu, Ramnit, ZeroAccess, and Metulji emerged in 2011. Kelihos and Chameleon malware hit the web world in 2012. Massive impact malware Windigo evolved in 2014 by 60% spam traffic. Mirai malware moved the world by huge impact in 2016. In late 2016, Leet IoT Botnet with massive DDoS attack created huge malicious impact on Internet and Internet of Things. Twitterverse, the enormous Twitter Botnet struck the world in 2017 (Touchette 2016). The evolution of malware since its origin is depicted in Fig. 5.2.

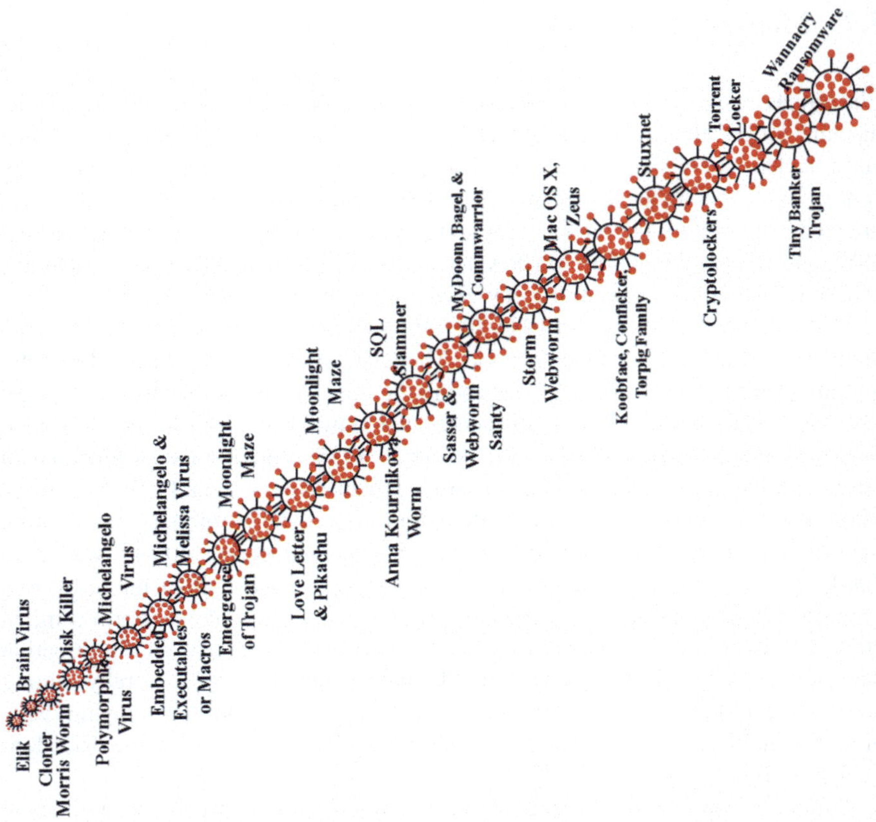

Fig. 5.2 Evolution of malware

5.3 Malicious Software

There are different kinds of malicious software which can be used in various techniques to perform the malicious operations. These include Virus, Worm, Trojan, Backdoors, Rootkit, Trojan-Spy, Trojan-Ransom, Bot, etc. are explained in this section.

5.3.1 Virus

A malicious program that is loaded into the computer and runs into the computer without the prior knowledge of the user is called the computer virus. Usually, viruses are the codes created by a human being with a malign intent. The virus has capability to replicate itself and that can be easily reproduced. This type of simple virus can also be very harmful as it consumes the memory and make the system slow. Extra harmful virus can bypass the security system and can spread across the networks. There are variety of viruses which cause huge damage to the companies in billions per year. There are various types of viruses such as macro virus, memory resident virus, direct action virus, web-scripting virus, Multipartite Virus, etc. as described in subsequent sections as follows.

5.3.1.1 Macro Virus

The macro virus infects the files created using the application that contains macros, i.e., doc, pps, xls, MDB files. They infect the files and document templates that are contained in the file. They are usually hidden in the documents shared in the network.

Example, Macro virus include Relax, bablas, Mellisa.A, 079M/Y2K.

5.3.1.2 Memory Resident Virus

The memory resident virus usually attaches itself inside the memory. It gets activated whenever the operating system runs and ultimately contaminates the opened files which are hidden in the RAM.

Example, CMJ, meve, randex, MRK lunky includes the memory resident virus.

5.3.1.3 Direct Action Virus

This virus mainly replicates on one occasion when they are executed. When a definite action is fulfilled, the virus will act accordingly infecting the file or directory or the folder specified in the AUTOEXEC.BAT. This type of viruses particularly reside in the hard disk's root directory but keep changing their location.

5.3.1.4 Web Scripting Virus

Websites with large population like social networking or email are targeted by web scripting virus. More common example of web scripting attack is the DDoS attack. There are two types of web scripting attacks such as non-persistent and persistent attack.

5.3.1.5 Multipartite Virus

Multipartite virus is a fast spreading virus which uses boot infectors to attack boot sector. It can act in multiple ways. At the same time, it can contaminate boot sector, executables and program files. This virus is more vulnerable for detection because of large number of places where the virus is spread in.

5.3.2 Worms

Worm is a self replicating malicious program that can propagate via networks without any human intervention. Normally, worms acquire considerable memory space and huge bandwidth so the affected computing systems and networks are overloaded and their performance slows down. Generally, computer worms are propagated through email attachments, web links, ICQ links, P2P networks, and network packets (Touchette 2016).

5.3.3 Trojans

Trojan malware generally looks like genuine software but actually it is used by malicious actors for getting access to target's computing system. By using some traps through social engineering, users are targeted by entering into their computing device. Once Trojan gets access to the system, it can steal sensitive credentials, spy on the computing system, and also can get backdoor access to the system. There are various forms of these malign actions including deleting data, blocking data, modifying data, copying data and disrupting the performance of computing system and networks (Saeed et al. 2013).

5.3.4 Backdoors

The backdoors are the types of Trojans which give users remote control over the infected computer. They enable the attacker to modify alter the data from the system and also monitor the transmission of the system which includes sending, receiving, and deleting files. Backdoor Trojans normally try to form a group of target computing machines to form a botnet or zombie network which is used for malign intents (Suarez-Tangil et al. 2019).

5.3.5 Exploit

Exploit malware takes benefit of software loopholes and security weaknesses for getting entry into the target device. Exploit is usually the first step towards a big attack. It uses a small malware payload called as shellcode for infecting target computing devices and infiltrates the company's sensitive information. Exploit kits are used for scanning the computing devices for the software loopholes and after detecting some loopholes, these kits install extra malware to increase the level of infection. Softwares such as Adobe Flash Player, Adobe Reader, Internet Explorer, Oracle Java, Sun Java, etc. are targeted by using exploit kits.

5.3.6 Rootkit

Rootkit is wicked computer software which hides itself deep inside the computing machine and keeps itself undetected. Rootkit can hide presence of worms, bot, and malware on the target machine. It is considered as a most dangerous form of malware. It usually keeps track of routine activities on the targeted machine, it scans the network traffic on the machine, installs the malicious programs without user's consent, it can also hijack the computing machine's resources, or it can make the machine the botnet slave. Rootkit malware consists of dropper, loader, and rootkit itself. Dropper installs the rootkit on the machine. Rootkit malware can enter the machine through a downloaded pdf or word document. Many types of Rootkits are there including kernel rootkit, hardware rootkit, virtualization rootkit, bootkit, memory rootkit, Necurs rootkit, etc. (Cucu 2017).

5.3.7 Trojan-Spy

Malicious program posing itself as a genuine application for propagation of spyware is known as Trojan-Spy malware. These are non-replicating kind of malwares. This Trojan-spy malware secretly installs spying programs or keylogger program on target machine. Trojan-spy can compromise user's sensitive credentials like logins and passwords, bank account number, credit card number, etc.

5.3.8 Trojan-Ransom

Trojan Ransom blocks or encrypts the data on the target machine. Because of this, the user is deprived of using that data. Infected targeted machine cannot function properly. The attackers demand a ransom amount to restore the compromised data. Trojan ransom malware can restrict the user from making use of applications which can confine Internet access, browser functionality, access to websites, access to operating system resources, user's actions in the operating system, etc.

5.3.9 Bots

Bots are basically the robots performing the programmed duty. Bots are of two types such as genuinely working bots and infected bots performing malicious actions giving complete control to the remote attacker. After getting infected, these bots are called as zombies. Huge number of computing machines are grouped together and within fraction of time, all of them can be infected. This type of infected network is called as Botnet. All the Botnet entities can be controlled at a time with the help of broadcasted harmful instructions. Internet connection to the user computing device should be secure one. Otherwise the infected device can easily become a part of the Botnet within very small time duration. Mass malicious actions are performed with these infected machines and whole cyber physical system can be shut down without knowing of the users.

Table 5.1 lists very recently observed and experienced malware attacks by the user. These malicious activities include Emonet, WannaCry, Kovter, Zeus, Dridex, IcedID, GhOst, Mirai, Nanocore and Pushdo. The malware attacks are increasing with new wireless technology (Center for Internet Security (CIS) Blog 2019).

Table 5.1 Latest malware activities

Sr. no.	Malware	Attack description
1	Emonet	Banking Trojans are dropped. Modular information stealer spreading throughout the network
2	WannaCry	Ransomware Cryptoworm infecting Windows computers by encrypting files on system hard drive, demand Bitcoin for decrypting of the files
3	Kovter	Click fraud malware and downloader. It hides itself into registry keys
4	ZeuS	Modular banking Trojan uses keystroke logging to compromise target's financial credentials
5	Dridex	Banking malware variant using malicious macros in Microsoft Office
6	IcedID	Modular banking Trojan malware that monitors target's online activities
7	GhOst	Remote access trojan (RAT) used for formation of backdoor into an infected device
8	Mirai	Botnet malware used to compromise IoT devices used to conduct mass DoS attack
9	NanoCore	Remote access trojan (RAT) spread through malspam as a malicious excel XLS worksheet
10	Pushdo	Botnet used to distribute Cutwail Spambot

5.4 Malware Investigation

In the Investigation of the malicious software, we have to understand the overall scenario before starting the investigation process. There are some ways to accomplish the malware investigation.

- To thoroughly examine the website that may be associated with the incident because they may be acting as an exploit that acts as an infection vector.
- Obtaining the IP addresses of the data of system involved in the incident because it is considered suspicious for hosting the malicious file that was dropped from the system.
- Keeping eye on IP addresses related to the infected parties in the blacklist to know if supplementary systems are not confronted.
- Performing programmed behavioral analysis of malware involved in the incident to get the common case of the distinctiveness.

5.5 Deception Methods

Deception technology is the cyber security defense mechanism against cyber attacks. Deception is capable of detect, analyze, and defend against real time attacks. The deception techniques are generally automated and pro-active in nature. In this section variety of deception techniques like anti-emulation, anti-online analysis, anti-hardware, anti-debugger, anti-disassemblers, and anti-tools are discussed.

Various deception methods are used in the analysis of the malware as stated below (Cobb and Lee 2014; Liţă et al. 2017):

5.5.1 Anti-emulation

VMWare or Virtual PCs are regularly scanned to see if any malware is running on this type of virtual machines. Timing based and artifacts based are the two techniques for emulation detection. In the timing based method, change in the time stamp is observed if the code emulation is going on. Under the artifacts based mechanism, malwares check the artifacts about virtual machines, related to files, network or device before getting speed for infection spreading.

5.5.2 Anti-online Analysis

Online malware analysis engines such as Anubis and Norman Sandbox are available for the cyber Forensic Analyst. Detailed reports about the malware actions can be obtained with the help of these engines. Online deception is carried out through information and communication technology (ICT). Because of technological advancements and digitalization, the scope for online deception has enhanced. Users are suffering from various types of social engineering attacks like spear phishing attacks.

5.5.3 Anti-hardware

Hardware deception is creating waves because of its vast and dangerous scope. Hardware components such as routers, biometric devices, IoT devices are compromised under this category. Deception decoy technology arranged throughout the network. This technique poses as if authentic ICT equipment and the hacker is attracted towards this decoy device. The attacker is tempted to get credentials from this decoy system and gets trapped into it. This activity is informed to the deception server.

5.5.4 Anti-debugger

If the malware is running in the debugger system, then anti-debugger deception technique is deployed. These techniques target the debugger of the computing system and exploit it to take control of the flow and execution of the malicious code. Debuggers can be programmed to detect such debugger deception activities and mitigate them while the program is running.

5.5.5 Anti-disassemblers

Actually, it is a trick used by disassemblers into showing incorrect code. Linear sweep and recursive traversal are the techniques used by disassemble for deception. This kind of malware attack genuine disassemblers and make them produce wrong disassemblies.

5.5.6 *Anti-tools*

The running malware can detect the tools used by the system to detect them. Once malware finds the tools, then it starts its deception activities. It tries to find the loopholes in the tools used, and use those loopholes against the tools only and tries another level of deception. Well known analysis engines can easily detect the tools.

5.6 Malware Detection and Analysis

In its initial stages, malware was a small trouble but now it has expanded and almost everywhere, malwares are present in some or the other form. Many automated monitoring technologies are available for malware detection. These techniques help in identifying and protecting against attacks like viruses, worms, Trojan horses, spyware, etc. The malware detection technology is helpful for protection of servers, gateways, user workstations, and mobile devices (Jadhav et al. 2016). There are two strategies that are worked out by professionals for performing the malware analysis mentioned as below:

5.6.1 *Static Analysis*

Without actually viewing the malicious code or instructions, static analysis can inspect the malware present on the system. It can quickly resolve if the file is infected or not and then also helps in producing malware signature. Technical details related to file name, MD5 checksums, hashes, file type, file size, and acknowledgment from antivirus tools can be obtained with the help of static analysis. The static analysis of the malware is done by analyzing the malware with the help of some tools. This analysis can be performed by using tools like virustotal. By the use of the virustotal, the complete analysis of the malware is done. It specifies that the analysis performed is done (Kim et al. 2018).

5.6.2 *Dynamic Analysis*

To observe the behavior of the malware present on the system, it is needed to run the malware, understand the way malware works and its strategies, and get the technical details to create malware signature. Dynamic analysis can provide information about domain names, P addresses, file path locations, registry keys, and extra files situated on the system. More often, automated sandboxes are used for dynamic analysis (Chouhan et al. 2014).

5.6.3 Mounted Analysis

Forensic image files are mounted on the logical drive of the machine under investigation where the infected files are present. Malware scanners are used against the mounted image. Malware scans are properly documented for future use. This mechanism speeds up the malware analysis. In mounted analysis, the infected file can be analysed at its native place and metadata can be easily examined (Lakhotia and Black 2017; Cao et al. 2013).

5.6.4 Booted Analysis

Exploitation of the basic input/output system (BIOS) boot process for launching the attack well before the operating system initialization is the big threat to cyber security of the computing device. Malware which runs during the earliest step of boot process is called as bootkit. These are persistent type of threats and are difficult to identify. There are two boot record namely master boot record (MBR) and volume boot record (VBR). To protect the machine from this kind of attacks, trustworthy collection of boot records is important. Also, behavioral analysis of boot records is necessary with static analysis. The ability of analyzing thousands of boot records within a fraction of time should be developed (Fisher and Davis 2018).

5.6.5 Network Analysis

Monitoring of network traffic to see if there is any change in the network behavior due to malware infection is a tedious process. Malwares use encrypted HTTPS protocol to bypass the network traffic analysis. By using machine learning techniques, the malware can be detected based on host addresses, time stamping, and volume information of the collected data (Prasse et al. 2017).

Table 5.2 lists the latest research on the anti-malware activities. Normally, the malware targets are mobile devices, computing machines, high and medium risk applications from enterprises, network traffic flows, etc. Researchers are working on various such issues and have come up with good detection, analysis and mitigating techniques as mentioned in the table. Upcoming technologies like artificial Intelligence and machine learning are the good solution provider platforms (Moser et al. 2007).

Table 5.2 Research on anti-malware activities

Sr. no.	Malware target	Anti-malware technique details	Reference papers
1	Over-privileged apps	Analysis of API call of applications	Moser et al. (2007)
2	Mobile malware	Permission based static analysis	Enck et al. (2009)
3	Distinguish between malign software and genuine software	Machine learning based analysis of permissions, API calls, and network addresses	Arp et al. (2014)
4	Distinguish between malign software and genuine software	DroidAPIMiner uses machine learning techniques and API calls of applications	Aafer et al. (2013)
5	First order analysis for high and medium risk applications and second order analysis for extraction of applications with obfuscating, encrypted or dynamic class loading techniques	RiskRanker—two level static analysis	Grace et al. (2012)
6	Library and system function calls	Two level static analysis	Schmidt et al. (2009)
7	Semantic signatures and model checking	Both static and dynamic analysis	Felt et al. (2011)
8	Mobile malware detection	Marvin	Lindorfer et al. (2015)
9	Import terms from Java codes are extracted by decompiling apk files to avoid obfuscation	MocDroid	Martin et al. (2016)
10	Mobile malware detection	Genetic programming (GP)	Le et al. (2014)
11	Network anomaly intrusion detection	Co-evolutionary computation based network behavior model	Ostaszewski et al. (2007)
12	Mobile malware detection and anti-malware development and application	Co-evolutionary computation for systems security	Sen et al. (2018)

5.7 Virtualization to Eliminate Malware

Virtualization is a step towards malware readiness and control. Virtualized computer units can be easily produced as well as destroyed with the movement of a switch. Sandboxing is the concept in virtualization in which an application or program is tricked and made to think that it is running on regular machine and monitoring its performance becomes easy. Virtualization sandbox does not need permanent management and requirement of the resources. When sandbox shows malware presence, the malware can be destroyed with the destroying of the virtual machine with the help of switching mechanism (TechAdvisory Editor 2017).

5.8 Summary

In this chapter, the analysis of different kind of malwares is done, and strategies for different technologies are discussed. Performance investigation of different types of malicious software like Virus, Trojan, Worms, Bots is discussed in this chapter. Different malware analysis strategies like mounted analysis, booted analysis, network analysis, and the virtualization malware are also elaborated. The anti-malware techniques such as anti-virus and firewalls must be used to protect against the malicious activities. The user awareness about the threats and vulnerabilities is going to help a lot to stay safe against attacks.

References

Aafer Y, Du W, Yin H (2013) DroidAPIMiner: mining API-level features for robust malware detection in android. In: Security and privacy in communication networks. Springer, pp 86–103

Arp D, Spreitzenbarth M, Hubner M, Gascon H, Rieck K (2014) Drebin: effective and explainable detection of android malware in your pocket. In: NDSS

Associated Press, ABC News (2017) The latest: FedEx confirms it hit by malware attack, 12 May 2017. http://abcnews.go.com/Technology/wireStory/latest-uks-health-service-hit-ransomware-attack-47372081

Cao Y et al (2013) Abstracting minimal security-relevant behaviors for malware analysis. J Comput Virol Hack Tech 9(4):193–204

Center for Internet Security (CIS) Blog (2019) Top 10 malware January 2019. https://www.cisecurity.org/blog/top-10-malware-january-2019/

Chouhan PK et al (2014) Network based malware detection within virtualized environments. In: Euro-Par 2014: parallel processing workshops. Springer International Publishing

Cobb S, Lee A (2014) Malware is called malicious for a reason: the risks of weaponizing code. In: 6th international conference on cyber conflict

Cucu P (2017) Rootkit—the (nearly) undetectable malware. Heimdal Security Blog. https://heimdalsecurity.com/blog/rootkit/

Enck W, Ongtang M, McDaniel P (2009) On lightweight mobile phone application certification. In: Proceedings of the 16th ACM conference on computer and communications security. ACM, pp 235–245

Felt AP, Chin E, Hanna S, Song D, Wagner D (2011) Android permissions demystified. In: Proceedings of the 18th ACM conference on computer and communications security. ACM, pp 627–638

Fisher R, Davis A (2018) BIOS boots what? Finding evil in boot code at scale. FIREEYE blog on threat research. https://www.fireeye.com/blog/threat-research/2018/08/bios-boots-what-finding-evil-in-boot-code-at-scale.html

Grace M, Zhou Y, Zhang Q, Zou S, Jiang X (2012) Risk ranker: scalable and accurate zero-day android malware detection. In: Proceedings of the 10th international conference on mobile systems, applications, and services. ACM, pp 281–294

Jadhav A, Vidyarthi D, Hemavathy M (2016) Evolution of evasive malwares: a survey. In: International conference on computational techniques in information and communication technologies (ICCTICT)

Kim CH, Kamundala KE, Kang S (2018) Efficiency-based comparison on malware detection techniques. In: International conference on platform technology and service (PlatCon)

Lakhotia A, Black P (2017) Mining malware secrets. In: 12th international conference on malicious and unwanted software (MALWARE)

Le TA, Chu TH, Nguyen QU, Nguyen XH (2014) Malware detection using genetic programming. In: 2014 seventh IEEE symposium on computational intelligence for security and defense applications (CISDA), Dec 2014, pp 1–6

Lindorfer M, Neugschwandtner M, Platzer C (2015) Marvin: efficient and comprehensive mobile app classification through static and dynamic analysis. In: 2015 IEEE 39th annual computer software and applications conference (COMPSAC), vol 2. IEEE, pp 422–433

Liţă CV, Cosovan D, Gavriluţ D (2017) Anti-emulation trends in modern packers: a survey on the evolution of anti-emulation techniques in UPA packers. J Comput Virol Hack Tech

Martin A, Menéndez HD, Camacho D (2016) MocDroid: multi-objective evolutionary classifier for android malware detection. In: SoftComputing, pp 1–11

Moser A, Kruegel C, Kirda E (2007) Limits of static analysis for malware detection. In: Twenty-third annual computer security applications conference. ACSAC, Dec 2007, pp 421–430

Ostaszewski M, Seredynski F, Bouvry P (2007) Coevolutionary-basedmechanisms for network anomaly detection. J Math Model Algorithms 6(3):411–431

Prasse P, Machlica L, Pevný T, Havelka J, Scheffer T (2017) Malware detection by analysing network traffic with neural networks. In: IEEE security and privacy workshops (SPW), USA

Saeed IA, Selamat A, Abuagoub AMA (2013) A survey on malware and malware detection systems. Int J Comput Appl (0975–8887) 67(16)

Schmidt A-D, Bye R, Schmidt H-G, Clausen J, Kiraz O, Yuksel KA, Camtepe SA, Albayrak S (2009) Static analysis of executables for collaborative malware detection on android. In: IEEE international conference on communications, 2009 (ICC '09). IEEE, pp 1–5

Sen S, Aydogan E, Aysan AI (2018) Coevolution of mobile malware and anti-malware. IEEE Trans Inf Forensics Sec. https://doi.org/10.1109/TIFS.2018.2824250

Suarez-Tangil G, Tapiador JE, Peris-Lopez P, Ribagorda A (2019) Evolution, detection and analysis of malware for smart devices. IEEE Commun Surv Tutor 16(2)

TechAdvisory Editor (2017) How virtualization roots out malware. https://www.techadvisory.org/2017/04/how-virtualization-roots-out-malware/

Touchette F (2016) The evolution of malware. Netw Secur 2016(1):11–14

Chapter 6
Copyright Infringement

American Chemical Society and Elsevier have filed a legal fight case against Research Gate. They blame Research Gate of "massive infringement of peer reviewed, published journal articles." According to them the networking site-Research Gate is illegally making research papers available to the users which are copyrighted (McKenzie 2018). "Media organization files copyright infringement case against Saint Peters Blog owner", the news on May 8, 2015, in Tampa Bay Business Journal, has created huge waves throughout the information and communication technology world (Huff 2015). Using media file contents for writing the blogs is a serious offence and comes under copyright infringement act. In the present day world, the number of Internet users has rapidly grown along with the growth of information and digital technology. With this development of the Internet and rapid increase in usage of digital multimedia, it has become very easy to generate unauthorized digital data.

6.1 Introduction

Keeping the intellectual property data safe against unauthorized access and usage has become very challenging in today's digital technology world. Therefore, it has created new challenges for protecting intellectual property rights (IPR). Due to the rapid growth of technology, there is an increase in the different cyber-crime attacks risk. The number of Internet users has grown rapidly with the growth of information and digital technology. Therefore, the problems such as copyright infringement and intellectual property rights (IPR) violation have emerged.

Copyright laws were prepared with the intention that the creator of an innovative work should have a right to decide how to reproduce the fruits from his talent, skill and hardwork. Copyrighting always protects the creator and his rights for the ownership for the same. Copyright is available for wide variety of talents like literary work, musical work, artistic work, cinematograph films, sound recording,

R. Prasad and V. Rohokale, *Cyber Security: The Lifeline of Information and Communication Technology*, Springer Series in Wireless Technology,
https://doi.org/10.1007/978-3-030-31703-4_6

computer programs, tables and compilations including databases. Copyright infringement occurs when someone other than the copyright holder or copyright owner uses or distributes information without permission. For example, illegally downloading music, image or cartoon from the website or a document is considered as copyright infringement. It is not easy to avoid copyright infringement. By using Technology, virtually anything can be copied quickly and nearly entirely (El-Wahed et al. 2007).

The chapter is organized as follows. Section 6.2 throws light on owner's rights and copyright infringement. Promising security solution for copyright infringement is discussed in Sect. 6.3. Section 6.4 elaborates classification of digital watermarking techniques. Robustness enabled digital watermarking procedures are illustrated in Sect. 6.5. Section 6.6 gives idea about state of the art security measures. Last Sect. 6.7 summarizes the chapter with insight to future research directions.

6.2 Owner's Rights and Copyright Infringement

An owner of copyright has certain rights to protect their work. An owner of a copyright can control on how their works are used by using these rights. If someone other than the copyright owner used these rights without any permission from the copyright holder or copyright owner, then we can say copyright infringement occurs (El-Wahed et al. 2007).

The rights of the copyright owner include

- Production of the work.
- Reproduction of the copies of work and distribution of these copies of work to public.
- Creation of new work, modification in the present work.
- Display the copies of work to public.
- Broadcasting the work to public. Examples sound recording playing, playing music, and showing video.

An infringement of copyright occurred when one uses any of these rights without the permission of an owner of copyright or copyright holder.

Copyright infringement occurs when someone other than copyright owner

- Reproduces the copies of work and distributes these copies without permission of copyright owner.
- Modifies the work made by copyright owner without permission.
- Displays the copies of work to public without copyright owner permission.
- Broadcasts the copyright holder's operates without any permission.

It is a significant challenge to protect our documents from piracy, i.e., violation or infringement of copyright.

6.2.1 *Examples of Copyright Infringement*

The following are the examples of copyright infringement

- Unauthorized download and share different video files, and games as well as download and share music MP3 files.
- Prohibited scanning the already published Photograph and use it.
- Putting more full-text articles on a web page and allowing all the Internet users to access that web page.
- Illegitimately using corporate logos.
- Provide unauthorized access to the licensed software without the permission of copyright holder.
- Using licensed images found on the Internet without permission.
- Download different videos found on the Internet and making a movie without permission of copyright owner.

6.3 Digital Watermarking: Promising Security Solution for Copyright Infringement

Digital watermarking can be considered as one of the strong security solutions for the phenomenon of infringement which will be clearer with the following description.

6.3.1 *Digital Watermarking*

Due to massive developments in computer and Internet technology, a multimedia data i.e. audio, video, image has found large applications in these days. Digital watermarking is one of the best solutions to prevent illegally copying, modifying and restructuring multimedia data. Security is the degree of protection from loss, damages, criminal activity and damages or resistance to harm. Nowadays, it's very important to provide security for the electronic documents. It's common to many people to share and transfer the multimedia data due to a rapid growth of Internet and networks technique. This multimedia data can be easily copied and modified; therefore need for copyright protection is growing.

Digital Watermarking is a process of embedding digital signals or pattern into a media object without affecting in any way the quality of the original file. While sending a message to the destination, it should be watermarked to protect this message from illegal copying, and it also provides the security. In digital watermark invisible information inserted into audio or video, image (Bhargava et al. 2012).

Digital Watermarking is nothing but hiding a digital signal message such as image, audio (song), video within the signal itself. The covering process has to be such that the modifications of the media are undetectable. Digital Watermarking is used to prevent or avoid illegal copying, redistributing and modifying multimedia data. By using digital watermarking, we can provide the copyright protection, covert communication, content identification and data authentication.

Primarily, the research of image watermarking is started, and then the watermarking is developed for audio. There are less watermarking is designed for audio. Since human auditory system (HAS) is more sensitive than the human visual system (HVS), it is very difficult to add watermark data in an audio file as compared to the image file (Dittmann et al. 2000).

The audio watermarking algorithms are not easy. Also, the available studies on audio watermarking are far less than that of video watermarking or image watermarking. Applying watermarks on audio data by using various algorithms is known as audio watermarking.

6.3.2 Overview of Digital Watermarking System

Digital Watermarking has been proposed as the technique for copyright protection of multimedia data. It is a process of embedding digital signals or pattern into a multimedia object without affecting in any way the quality of the original file. While sending a message to a destination, it should be watermarked to protect this message from illegal copying, and it also provides the security. In digital watermarking process, hidden information is covered inside audio, video, or image (Zhong and Chen 2011).

The block diagram of a digital watermarking system is shown in Fig. 6.1.

A digital watermarking system has three blocks:

(1) Embedding
(2) Attack
(3) Detection/Extraction

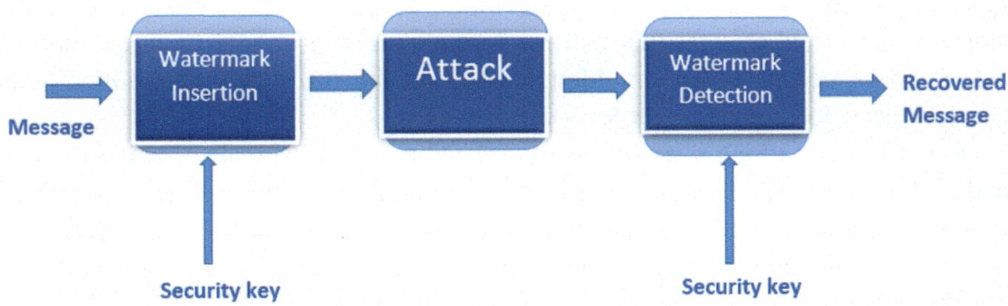

Fig. 6.1 Digital watermarking system

The security key and data to be embedded (message) are the inputs to the watermark insertion block, and it produces the watermarked signal.

- The Watermarked digital signal is transmitted or stored, usually sent to another person. If this person makes a modification, this is called an "Attack."
- Then this distorted watermarked signal is applied to watermark detection block with the key. Watermark detection block gives the recovered message as an output as shown in Fig. 6.1.

6.3.3 Properties of Digital Watermark

Properties of digital watermark like Robustness, Imperceptibility, Capacity and data payload, Security and Computational cost are explained in the following paragraph.

Robustness Robustness is nothing but the watermark embedded in data can withstand non-malicious distortions (Bhargava et al. 2012). Watermark data should be robust so that it should not be recognized by unauthorized distributors and watermark data should resist many attacks such as filtering, compression and filtering with compression, noise addition, cropping, resampling, Re-quantization.

Imperceptibility The watermark should be undetectable, impossible to hear to human ears, and should only be detected by an authorized person. Imperceptibility is nothing but hiding a watermark so that the watermark should not affect the quality of the original signal.

Capacity and Data Payload Capacity is nothing but the number of bits that can be embedded in one second of the host signal. If multiple watermarks are embedded into data, then the watermarking capacity of the image is the sum of all individual watermarks data payload.

Security Security of a watermark is the ability that authorized person should only extract the watermark. Security resists malicious attacks (Dittmann et al. 2000). Security provides watermark access only to the authorized parties.

Computational Cost If the watermarking method is less complicated then the computational cost of the watermark should reduce. If computing resources such as hardware and software required for the watermarking method are with high sophisticated algorithms, then it incurs the more computational cost. If the watermarking method is simple, then the limited resource will require and also computational cost should be less. Thus, the computational cost and complexity of watermark methods are directly proportional to each other and vice versa.

We need to make a trade-off between all these requirements/properties because they are often contradictory with each other.

For example, increasing data rate or capacity in watermark system results in the reduction of the robustness against attacks and also degradation of the quality of the watermarked signal as depicted in Fig. 6.2.

6.4 Classification of Digital Watermarking Techniques

To control the copyright of digital material, a digital watermarking technique is used. Digital watermarking techniques can be classified in various ways. The classification chart of digital watermarking is depicted in Fig. 6.3. Digital watermarking techniques can be categorized according to Watermark Embedding Domain, Based on the type of Documents, Rest on Perceptivity and depending on Use at. The classification of digital watermarking is shown in the figure given below. According to working domain there are two types of watermarking technique, Time domain and Transfer domain (Bhargava et al. 2012). According to the multimedia data used watermarking is further divided into the image watermarking, text watermarking, audio watermarking and video watermarking techniques. Based on perceptivity, watermarking techniques are classified into visible watermarking and invisible watermarking. There are two types of digital watermarking depending on use at are Source based watermarking and Destination based watermarking techniques (Dittmann et al. 2000). Also, Table 6.1 shows classification of watermarking techniques.

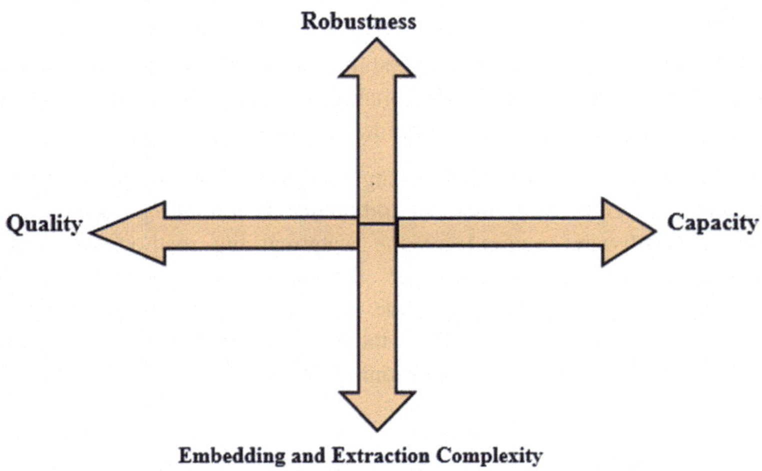

Fig. 6.2 Properties trade-off

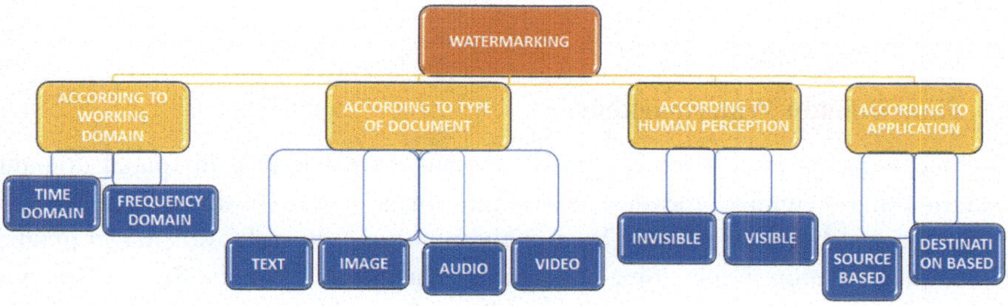

Fig. 6.3 Digital watermarking classification chart

Table 6.1 Classification of watermarking techniques

Classification	Content
Watermark embedding domain	Spatial domain, transform domain
Multimedia type	Text, image, audio, video
Human perceptibility to watermark	Visible, invisible
Use at	Source based, destination based

6.4.1 According to Watermark Embedding Domain

The first criterion of digital watermarking techniques classification is according to watermark embedding domain (Bhargava et al. 2012). Digital watermarking classification based on embedding domain is shown in Fig. 6.4. Using this criterion, techniques can be classified into two types:

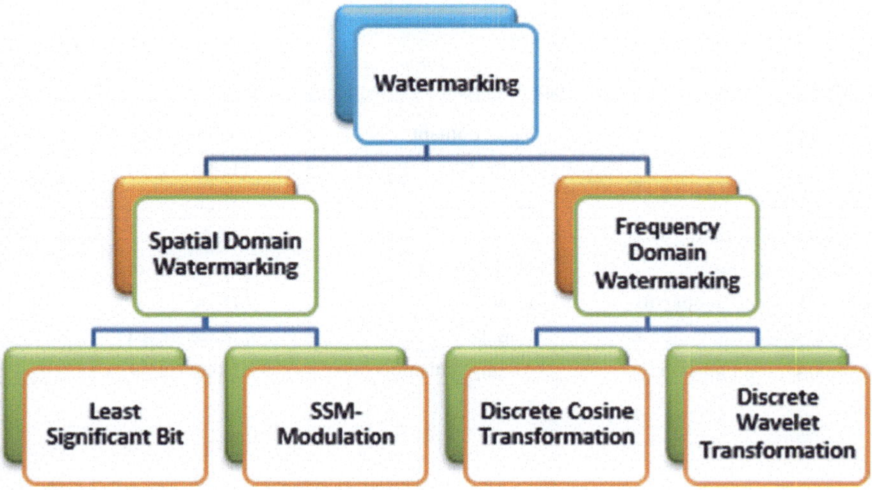

Fig. 6.4 Domain based watermarking classification

(1) Time Domain
(2) Transform Domain

(1) **Time Domain Watermarking**

This technique is also called as spatial domain watermarking. In spatial domain watermarking technique, message inserts into image pixels. In this spatial domain method, a watermark is inserted by directly modifying the values of image pixels. The implementation in this domain is simple. Watermarking in the Spatial Domain is based on Least Significant Bit (LSB). But information hiding capacity for this algorithm is low. This algorithm depends on the direct manipulation of image pixels. In this technique, the raw data is directly inserted with the multimedia data as an image. Some spatial domain watermarking algorithms are LSB algorithm, SSM Modulation based technique. In this technique, data is hidden without any transformation and. Therefore, it can be destroyed easily.

(2) **Transform Domain Watermarking**

This technique is also known as Frequency domain watermarking. Information hiding capacity for this algorithm is more as compare to time domain algorithm. In this transform domain method, the watermark is implanted by modifying the transform coefficients and not the image pixels values. Examples: Discrete Wavelet Transform (DWT), Spread Spectrum Watermarking, Discrete Cosine Transform (DCT), Discrete Fourier Transform (DFT). This algorithm gives more robustness against watermarking attacks. Table 6.2 gives the comparison of Time domain watermarking technique and Frequency domain watermarking technique.

In time domain method (Spatial domain) data is embedded without any transformation and, therefore, it can be destroyed easily. Also spatial domain method has poor robustness, therefore, we will focus on Transform based (Frequency domain based) watermarking technique only.

Table 6.2 Comparison between spatial domain and frequency domain

Parameters	Time domain	Transform domain
Capacity	High	Low
Computation cost	Low	High
Computational time	Less	More
Perceptual quality	High control	Low control
Computational complexity	Low	High
Robustness	Fragile	More robust
Example of application	Mainly authentication	Copy rights

6.4.2 *Based on Type of Document*

The second criterion of digital watermarking techniques classification is based on a type of document. Using this criterion, techniques can be classified into three types:

- Image watermarking
- Video watermarking
- Audio watermarking

Several digital watermarking techniques are proposed which includes watermarking for images, audio, and video.

Image Watermarking This adds watermark to an image. It is used to hide the information in the image and this information later detects and extract for author's ownership.

Video Watermarking Here watermark is applied to individual frames of a video. Video Watermarking is nothing but the extension of image watermarking. Video watermarking imposes real or near real-time watermarking system.

Audio Watermarking Digital audio watermarking is used for protecting audio files against illegal copying. It embeds the watermark to audio files such music, MP3.

6.4.3 *Rest on Perceptivity*

The third criterion of digital watermarking techniques classification is based on Perceptivity. Using this criterion, techniques can be classified into two types:

- Visible Watermarking
- Invisible Watermarking

(1) **Visible Watermarking**

In Visible watermark technique, digital data watermarking is noticeable like stamping a watermark on paper, like HBO, television channels, whose logo is clearly placed over on the corner of the TV picture. Watermark embedded into visual content such a way that it can be seen.

(2) **Invisible Watermarking**

In this technique, in the digital media such as image, audio, video, we can insert the raw data or secret information which is not visible and that cannot be seen. But can be extracted by using specific software and right process. This watermarking is Imperceptible. These watermarks cannot be detected by just viewing the digital content. Invisible watermark is also called as transparent watermarks.

6.4.4 Depending on Use at

The last criterion of digital watermarking techniques classification depends on use at. Using this criterion, techniques can be classified into two types source based and destination based. Source-based watermark is advantageous for ownership authentication where a unique watermark implying the owner is introduced to all the copies of a particular image being distributed.

Destination based watermark where each distributed copy gets a unique watermark implying the specific purchaser.

6.5 Digital Watermarking Based on Robustness

The following sections elaborate the concepts of digital watermarking depending upon the robustness, the robust watermarking scheme, fragile watermarking scheme and semi-fragile watermarking scheme. The Internet provides access to a lot of useful information. Therefore over a past few years, the growth of use these Internet services has been widely increases. Technology has no limits today. The rapid development of new technology has changed the way we work. The Internet has a lot of advantages, but there are also some disadvantages of using it. With the rapid development of new technology, it becomes very easy to perform illegal copying of the multimedia data such as image, audio and video. Therefore nowadays, to provide security for such attacks is the very important issue. In the last years, digital watermarking technology has emerged which provides the copyright protection for such multimedia data in the form of image, audio and video (Zhong and Chen 2011).

Researchers of digital watermarking focus on the classification of watermarking based on the Robustness. Robustness is nothing but the watermark embedded in data can withstand non-malicious distortions. Watermark data should be robust so that unauthorized distributors should not eliminate it. Robustness is nothing but the ability to resist many attacks (Sang and Alam 2008).

Digital Watermarking techniques can be classified based on the robustness into three main categories:

1. Robust
2. Fragile
3. Semi-fragile.

6.5.1 Robust Watermarking Scheme

Robust watermarks are used for copyright protection. It is also used for tracking the content. Robust watermark can prove ownership claims. Robust watermarks are designed to resist many attacks which attempt to destroy or remove the watermark (Sang and Alam 2008). In this watermarking technique modification to the watermarked content will not affect the watermark. It resists a designated class of transformations. Robust watermarking has ability to withstand accidental or malicious manipulations of the watermarked document. Robust watermark is useful for copy control, fingerprinting as well as ownership assertion.

6.5.2 Fragile Watermarking Scheme

In this technique, if watermarked content is changed or modified then watermark gets destroyed. Fragile watermarks are also used for copyright protection (Sang and Alam 2008). It is used for illegal alteration of a document prevention/detection. This watermarking technique is very sensitive. Fragile watermarking is applied to multimedia content authentication. Fragile watermark is used for integrity proof. If watermarked document is manipulated somewhat then fragile watermarks gets change.

6.5.3 Semi-fragile Watermarking Scheme

Semi-fragile watermark is the combination of fragile watermark properties and robust watermark properties. It is used for the content authentication. The semi-fragile watermark can be designed to tolerate legitimate changes while highlighting intentional distortions. It is also used for tamper detection. Semi-fragile was watermarking having the ability to tolerate some degree of modification done in an image such as the addition of quantization noise from lossy compression. This mechanism is capable of localizing regions of the image that have tampered and distinguishing them from regions that are still authentic.

6.6 State of the Art Security Measures

It is very critical to limit copyright infringement. To control the copyright of digital material, the digital watermarking technique is used. Digital Watermarking is nothing but hiding a digital signal message such as image, audio (song), video within the signal itself. The hiding process has to be such that the modifications of the media are imperceptible.

Primarily, the research of image watermarking is started, and then the watermarking is developed for audio. There are less watermarking is developed for audio. There are less watermarking techniques are proposed for audio compared to the image/video. State of the art security measures for copyright infringement is explained in Table 6.3. The different methodology is used for each paper, and accordingly advantages and disadvantages of each method are depicted in given Table 6.3 shown.

In High Capacity Digital Audio Reversible Watermarking paper Integer transforms reversible image watermarking scheme is used, and this method has High SSNR, but this approach is not annoying for listening. A Dual Digital Audio Watermarking Algorithm Based on LWT has the advantage to resist many attacks.

An Audio Digital Watermarking Algorithm against A/D and D/A Conversions Based on DCT domain has high robustness and SNR. In Robust Patchwork-Based Embedding and Decoding Scheme for Digital Audio Watermarking paper, DCT algorithm is used. It gives high imperceptibility. Reversible Watermarking for Digital Audio-Based on Cochlear Delay Characteristics method includes some distortion. Other methods for the security of audio watermarking such as based on Fast Fourier Transformation (FFT), DWT and DCT, Nonlinear amplitude limiting computer-generated hologram, Double DCT transform, Artificial intelligent technique, Matlab Technique, DWT, Neural Networks gives high robustness. But WAV Format Audio Digital Watermarking Algorithm Based on HAS having Limit in small number formats.

6.7 Summary

Copyright infringement occurs when someone other than the copyright holder use or distribute information without permission. For example, illegally downloading music, image or cartoon from a website or a document comes under it. It is not easy to avoid copyright infringement. By using technology, virtually anything can be copied easily and nearly perfectly. To control the copyright of digital material, a digital watermarking technique is used. Watermarking is a process in which additional data is embedded along with the underlying data i.e. audio, images, and video. The future research directions for audio watermarking techniques are given in Table 6.4. Available studies on audio watermarking are far less than that of video watermarking or image watermarking.

Applying watermarks on audio data by using various algorithms is known as audio watermarking. As the human ear is more sensitive, it can detect an even small amount of embedded noise. Therefore, these audio watermarking algorithms are not easy. In time domain method (Spatial domain) data is embedded without any transformation and. Therefore, it can be destroyed easily. Also spatial domain approach has reduced robustness. Therefore, we will focus on Transform based (Frequency domain based) watermarking technique only. Among all the frequency domain techniques, such as spread-spectrum watermarking (SSW), Amplitude Modification, Replica Method, Dither watermarking, the main difficulty is that data embedding will cause large amounts of auditory noise. In our work, we propose to synthesize (reduced) noises completely and also reduced the quality degradation of

Table 6.3 State of the art security measures for copyright infringement

Sr. no.	Paper title	Methodology used	Advantages	Disadvantages
1	High capacity digital audio reversible watermarking	Integer transform reversible image watermarking scheme	High SSNR	Not annoying for listening
2	A dual digital audio watermarking algorithm based on LWT	LWT	Resist many attacks, such as: Gaussian noise, salt and pepper noise, low pass filtering	
3	Visualization of digital audio watermarking based on the dyadic wavelet transform	Two-dimensional dyadic wavelet transform (DYWT)	Better quality, robust	Low SSNR
4	An audio digital watermarking algorithm against A/D And D/A conversions based on DCT domain	DCT domain	Very good robustness, SNR is greater	
5	Robust patchwork-based embedding and decoding scheme for digital audio watermarking	Novel patchwork-based embedding and decoding scheme (DCT)	High imperceptibility, high robustness	
6	Reversible watermarking for digital audio based on cochlear delay characteristics	Cochlear delay (CD)		Distortion
7	Robust FFT based watermarking scheme for copyright protection of digital audio data	Fast fourier transformation (FFT)	Higher SNR, resists various attacks such as noise addition, cropping, resampling, re-quantization, and MP3 compression	
8	Large capacity digital audio watermarking algorithm based on DWT and DCT	DWT and DCT	Simple, easy to implement, imperceptibility and robustness is very good, large capacity	
9	Digital audio watermarking based on holographic nonlinear limiter	Nonlinear amplitude limiting computer-generated hologram	Good security, more robustness, lossy compression, re-sampling, low-pass filter, noise	

(continued)

Table 6.3 (continued)

Sr. no.	Paper title	Methodology used	Advantages	Disadvantages
10	An audio digital watermarking algorithm transmitted via air channel in double DCT domain	Double DCT transform	Good hidden effect, very good robustness, resist noise	
11	WAV format audio digital watermarking algorithm based on HAS	HAS	Good imperceptibility, robustness and strong resistance of usurping the ability of the copyright	Limit in small number formats
12	An efficient digital audio watermarking scheme based on genetic algorithm	Artificial intelligent technique	Improve quality and robustness, robust against various signals processing such as re-sampling, re-quantization, low-pass filtering, noise adding, random cropping, MPEG 1 layer III compression	
13	Effective watermarking of digital audio and image using Matlab technique	Matlab technique	Security and robustness	
14	A new algorithm for digital audio watermarking based on DWT	DWT	Good imperceptibility and robustness, better imperceptibility	
15	Digital audio watermarking algorithm based on neural networks	Neural networks	High robustness, transparency, low computational complexity	

embedded signals to improve security. Also, we propose to reduce the complexity of the system such that computational cost gets minimized.

Cyber Security is a very critical issue for the forthcoming fifth generation mobile communication where huge data will be generated. To save that data from various cyber-attacks is the real challenge. A negligible amount of attention has been given towards the security and reliability of the audio data which contains very important information. Development of robust and secure audio watermarking techniques is the need of the forthcoming converged internet age.

Table 6.4 Future research directions

Paper no.	Method used	Robustness	Imperceptibility	Security	Capacity and data payload	Computational cost and complexity reduced	Noise	Quality
Choi and Pun (2013)	Integer transform reversible image watermarking scheme			Not concentrated on security issues	✓	Not addressed uptill now	✓	
Xuesong et al. (2012)	LWT	✓						✓
Minamoto et al. (2012)	Two-dimensional dyadic wavelet transform (DYWT)	✓						✓
Guo et al. (2012)	DCT domain	✓						✓
Natgunanathan et al. (2012)	Novel patchwork-based embedding and decoding scheme (DCT transform used)	✓	✓					
Unoki and Miyauchi (2011)	Cochlear delay (CD)						✓	
Dhar and Echizen (2011)	Fast fourier transformation (FFT)			✓				
Ren and Li (2011)	DWT and DCT	✓	✓		✓			
Chen et al. (2011)	Nonlinear amplitude limiting computer-generated hologram (CGH)	✓		✓				
Chang et al. (2011)	Double DCT transform	✓		NOT concentrated on security issues				
Cai and Chen (2011)	HAS	✓	✓		✓			
Kumsawat (2010)	Artificial intelligent technique	✓						✓
Subbarayan and Karthick	Matlab technique	✓			✓			

(continued)

Table 6.4 (continued)

Paper no.	Method used	Robustness	Imperceptibility	Security	Capacity and data payload	Computational cost and complexity reduced	Noise	Quality
Ramanathan (2009)								
Meng et al. (2009)	DWT	✓	✓					
Hu et al. (2008)	Neural networks	✓				✓		

References

Bhargava N, Sharma MM, Garhwal AS, Mathuria M (2012) Digital image authentication system based on digital watermarking. In: International conference on radar, communication and computing (ICRCC), pp 185–189

Cai Q, Chen Y (2011) WAV format audio digital watermarking algorithm based on HAS. In: International conference on control, automation and systems engineering (CASE), pp 1–4

Chang D, Yang W, Huang Q, Guo W, Zhao Y (2011) An audio digital watermarking algorithm transmitted via air channel in double DCT domain. In: International conference on multimedia technology (ICMT), pp 2926–2930

Chen D-Q, Gu J-H, Zhou H (2011) Digital audio watermarking based on holographic nonlinear limiter. In: International conference on electronics and optoelectronics (ICEOE 2011), vol 2, pp 91–94

Choi K-C, Pun C-M (2013) High capacity digital audio reversible watermarking. In: IEEE CYBERNETICSCOM, pp 72–75

Dhar PK, Echizen I (2011) Robust FFT based watermarking scheme for copyright protection of digital audio data. In: Seventh international conference on intelligent information hiding and multimedia signal processing, pp 181–184

Dittmann J, Mukherjee A, Steinebach M (2000) Media-independent watermarking classification and the need for combining digital video and audio watermarking for media authentication. In: International conference on information technology: coding and computing, pp 62–67

El-Wahed S, Elfatatry A, Abougabal MS (2007) A new look at software plagiarism investigation and copyright infringement. In: ITI 5th international conference on information and communications technology, pp 315–318

Guo Q, Zhao Y, Cheng P, Wang F (2012) An audio digital watermarking algorithm against A/D and D/A conversions based on DCT domain. In: 2nd international conference on consumer electronics, communications and networks (CECNet), pp 871–876

Hu J, Qiu X-M, He D-T (2008) Digital audio watermarking algorithm based on neural networks. In: International conference on apperceiving computing and intelligence analysis, pp 89–92

Huff P (2015) Media organization files copyright infringement case against Saint Peters Blog owner. Tampa Bay Bus J. http://www.bizjournals.com/tampabay/news/2015/05/08/media-organization-files-copyright-infringement.html

Kumsawat P (2010) An efficient digital audio watermarking scheme based on genetic algorithm. In: International symposium on communications and information technologies (ISCIT), pp 481–485

McKenzie L (2018) Publishers escalate legal battle against research gate. Inside higher ED blog. https://www.insidehighered.com/news/2018/10/04/publishers-accuse-researchgate-mass-copyright-infringement

Meng L, Yang S, Jiang Q (2009) A new algorithm for digital audio watermarking based on DWT. Glob Congr Intell Syst 4:229–233

Minamoto T, Ogata Y, Sawai M (2012) Visualization of digital audio watermarking based on the dyadic wavelet transform. In: Ninth international conference on information technology—new generations, pp 640–645

Natgunanathan I, Xiang Y, Rong Y (2012) Robust patchwork-based embedding and decoding scheme for digital audio watermarking. IEEE Trans Audio Speech Lang Process 20(8): 2232–2239

Ren K, Li H (2011) Large capacity digital audio watermarking algorithm based on DWT and DCT. In: International conference on mechatronic science, electric engineering and computer, pp 1765–1768

Sang J, Alam MS (2008) Fragility and robustness of binary-phase-only-filter-based fragile/semi fragile digital image watermarking. IEEE Trans Instrum Measur 57(3)

Subbarayan S, Karthick Ramanathan S (2009) Effective watermarking of digital audio and image using Matlab technique. In: Second international conference on machine vision, pp 317–319

Unoki M, Miyauchi R (2011) Reversible watermarking for digital audio based on cochlear delay characteristics. In: Seventh international conference on intelligent information hiding and multimedia signal processing, pp 314–317

Xuesong C, Haiman C, Fenglei W (2012) A dual digital audio watermarking algorithm based on LWT. In: Proceedings of 2012 IEEE international conference on measurement, information and control (MIC), vol 2, pp 721–725

Zhong D, Chen C (2011) The study of digital watermarking system for the protection of multimedia courseware. In: IEEE 2nd international conference on computing, control and industrial engineering (CCIE), pp 304–307

Chapter 7
Cyber Forensics

Cyber Forensics is a science wherein the scientific examination and analysis of digital evidence are done so that the information obtained is put up into the presentable format which can be used as appropriate proof in the court of law. Cyber forensics has very much vast context associated with digital forensics including homeland security, information security, corporate espionage, economic spying, white collar crime, child pornography, traditional crime, incident response, employee monitoring, privacy issues, etc. According to Dave McCurdy, "Security is not a one-time activity but rather a continuous, risk-managed process". This is very true in the sense that none is robustly designed and well-proven security algorithm is all the time secure. With the global connectivity ensured with upcoming fifth generation mobile communication with the smart machine to machine communication, cyber security solutions should also be too quick and safe.

7.1 Introduction

Cyber Forensics has very old and exciting history with origin from United States Military and Intelligence Agencies. Cyber Crime Legislation was laid down during 1970s in U.S. Law enforcement investigative (LE) units were started around 1980s. By those times, the cyber-crime investigations were focused on drug, murder and child pornography crimes. International LE meeting was held in the early 1990s and it was the birth place of the uniform policies for international legislation for cyber-crimes. During 1990s, first International Organization on Digital Evidence (IOCE) was held in Europe following the Scientific Working Group on Digital Evidence (SWGDE) through a collaborative effort of the Federal Crime Laboratory Directors in 1998. First Regional Computer Forensics Laboratory (RCFL) was formed in Russia in 2000. It followed by Convention on Cyber Crime (COE) and

R. Prasad and V. Rohokale, *Cyber Security: The Lifeline of Information and Communication Technology*, Springer Series in Wireless Technology,
https://doi.org/10.1007/978-3-030-31703-4_7

first Digital Forensic Research Workshop (DFRWS) was held in New York in 2001. A lot of attention was paid to cyber-crime attacks and cyber security after that. Nowadays, Cyber security is a lifeline for whole information and communication (ICT) world.

Cyber Forensics activities mainly consist of safe collection of computer data, suspect data identification, analysis of suspect data without alterations to find out origin and useful content, presentation of the acquired information to the court of law and ultimately, application of a derived country's cyber security law to the computer practice and services associated with business industry. These activities are elaborated in Fig. 7.1.

The organization of the chapter is as follows. Section 7.2 discusses cyber-crime threat scenarios. Threat management is elaborated in Sect. 7.3. Section 7.4 throws light on incident response and cyber forensics. Network forensics is described in Sect. 7.5. Section 7.6 gives information about cloud forensics. Memory Forensics is elucidated in Sect. 7.7. Evidence collection and analysis is deliberated in Sect. 7.8. Section 7.9 explains data acquisition mechanism. Standardization activities related to cyber forensics are discussed in Sect. 7.10. The chapter is summarized in Sect. 7.11.

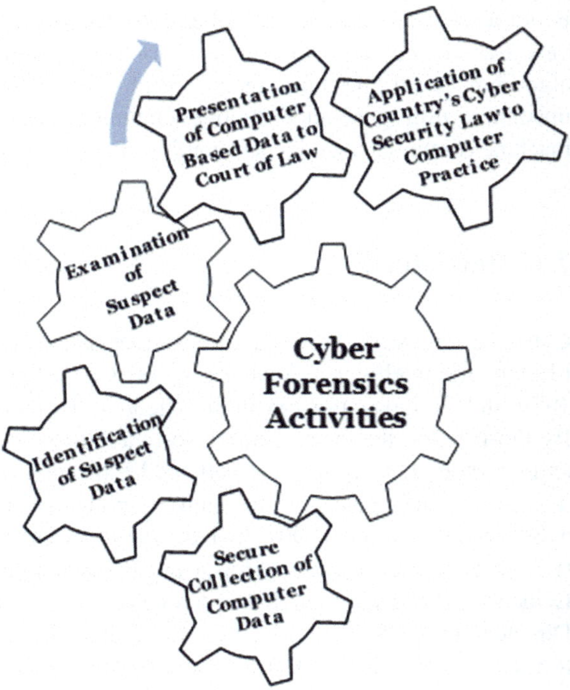

Fig. 7.1 Various cyber forensics activities

7.2 Cyber Crime—Threat Scenarios

One country can improve demeanor criminal attacks on the foreign because cyberspace has no boundaries as such. There are a lot of differences between the various countries and their laws. So it becomes tough for keeping the track of such malicious activities and judging these types of crimes and sentences the attackers. This kind of circumstances produces massive complications in the cyber prosecution process. The main offices are then located in the countries where the risks are very fewer.

Cyber criminals and mal-practitioners very cleverly advertise their services for promoting their business. It becomes tough for the user to think them as of malicious programs. The countries which have weak cyber security arrangements are the feasts for these cyber attackers. Online cheats, cyber espionage, copyright infringement, small fault sophisticated schemes, intelligence lacking in the trust management mechanisms are some of the common malpractices which user has to face in his/her everyday life. White collar criminals always refine their strategies to attract the users and easily cheat them. It is the very likely thing that for financial transactions, banks have started using one-time-use passwords (OTPS), grid system and sending codes to user mobiles. This has increased the security level of such businesses (Paganini 2012).

There are various ways by which users are cheated in practical life. Table 7.1 shows the typical attack incidences, their implications for users and how to avoid being the victim of such malware practices (Threat Encyclopedia 2013). Subsequent section gives idea about threat management.

7.3 Threat Management

Cyber security threat management is the very crucial task. Threat management needs to be done very intelligently with consideration of users as well as services and service providers as well. It involves almost everything including cyberspace user awareness, user privilege management, removable media control, ICT system monitoring, secure ICT configuration, malware protection, remote or mobile working, incident management, etc. as depicted in Fig. 7.2. Incident response and cyber forensics is described in next section.

7.4 Incident Response and Cyber Forensics

Cybercrime is a very critical issue for the whole world because it affects international cooperation, global coordination of law regulation and future amendments in these aspects. The international community should be formed which will work in

Table 7.1 Common scenario of cyber victims

Sr. No.	Malicious activity	User implications	Safety precautions
1.	Ransomware	Locks user's files or system for money. System file hostage is captured for monetary benefits	(1) Always take system back up (2) Stay away from downloading the data from unverified sources
2.	Online financial theft	User bank account savings are drastically reduced or transferred to unknown accounts	(1) Periodically change the passwords of online bank transaction accounts (2) Bookmark the online banking webpages
3.	Social networking profile spamming	Accidently user signs into fake login websites or clicks on malicious links which results into spamming	(1) Avoid clicking every link while browsing which appears fascinating
4.	Premium service abusing	User receives big fat mobile phone bills without that much usage	(1) Avoid unauthorized mobile security provisions (2) Check all the apps on mobile with their permissions
5.	Survey scam	While downloading a desired video or software, unknowingly user gets into a survey scam through social engineering and Black hat SEO	(1) Use reliable websites (2) Don't try various unknown search engines
6.	Mobile malware	Smartphone device battery drains very fast. It doesn't last long as usual	(1) Avoid downloading from unauthorized apps (2) Make use of mobile security solution which can block such websites
7.	Socially engineered spam attack	User's system becomes slow and unresponsive after opening a file attachment or visiting a link in their mailboxes	(1) Verify with the sender about suspicious mails (2) Delete all unfamiliar emails from the inbox

collaboration with industry and academia to address cybercrime issues and to find appropriate solutions for the same. There is need of a robust and balanced universal strategy for the cyber warfare. A lot of research is required to find solutions for these kinds of malign cybercrime activities (Sekgwathe and Talib 2012).

A number of major decisions and actions are involved while providing an efficient computer security incident response competence. First step is to form a group. Then team decides the work and services provision strategy regarding incident response plan, policy, and process formation are a critical part of establishing a team (Scarfone et al. 2008; West-Brown et al. 2003).

Forensics task force team of incident handlers should be equipped with good knowledge of legal aspects, guidelines, procedures, tools, techniques, anti-forensics instruments handling, and technologies related to it. Information security expert

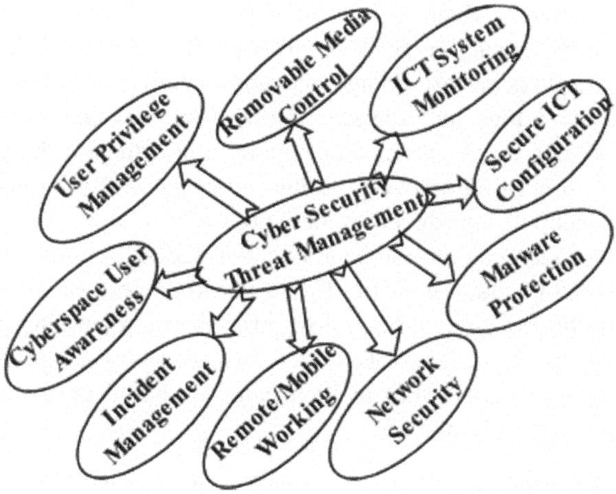

Fig. 7.2 Cyber security threat management

should be a part of team and he should have awareness about most commonly used OSs, file systems, applications, and network protocols within the organization. This kind of preparedness usually increases the speed and efficiency of the process.

Almost all the members of the incident handling team should have sufficient knowledge regarding all the aspects of the whole process so that in the absence of a team member, the teamwork will not be affected. Acquaintance up-gradation about new skills, forensics technologies, techniques, procedures and tools add value to the process (Kent et al. 2006). Following section gives information about network forensics.

7.5 Network Forensics

Once the network host is compromised by the malicious attacker, then the attacker can erase all the log information files over there. Under such circumstances, network traffic based evidence is the only evidence available for Forensics analysis. Network Forensics is the science and art of capture, record, and analysis of network events which can assist in finding out the origin of the malign activity or attack on the security system. Two simple practices are considered in the Network Forensics, leading one is related to abnormal network traffic monitoring and intrusion identification. The other one is associated with law enforcement which consists of rearranging the transferred files, keyword probing, and personal communication analyzing. Research work in (Raftopoulos and Dimitropoulos 2013) contains information about complex experiment conduction and its analysis which gives us real insight into Forensics analysis processes. Normally, security sources, Snort alerts, reconnaissance and vulnerability scanners, blacklists, and a search engine, to manually investigate these incidents actually provide necessary information to the researchers.

Recent advancements in wireless technology and its applications have posed more attacks against enterprises organizations and user computing devices. Attackers are becoming very much intelligent by removing attack tresses including system logs and registry information. Because of this attack analysis becomes difficult due to absence of important information. Industries and organizations have started using network traffic information for attack analysis. Cyber Blackbox concept is introduced for network traffic analysis (Choi et al. 2016). Cyber blackbox manager can provide various interfaces to users for searching information and creating attack scenario. Cyber blackbox performs tasks such as storage of network traffic and generation of flow information, rebuilding of the file sent by network and generation of metadata, data management by safeguarding the integrity of collected data, perform cyber incident analysis based on available data and sharing of the incident information.

7.6 Cloud Forensics

Cloud forensics is the amalgamation of cloud computing and digital forensics. In the Internet era, the users, enterprises, and organizations are always connected and almost all the data is stored on the cloud. Variety of cyber crimes are intended towards cloud data and digital forensics helps to inspect these attacks. In the research work of (Saibharath and Geethakumari 2015), the authors have proposed a mechanism of remote evidence collection and pre-processing framework by using Struts and Hadoop system. Pull model performs the task of collection of virtual machine disk images, logs, and other information when triggered by user. Cloud node sends the network captured data to the Hadoop File System. Cross drive analysis including clustering and correlation of logs and VM disk images is done by Mahout and Weka. Time required for forensics investigation is considerably reduced with this technique.

7.7 Memory Forensics

Forensics analysis of volatile data in the computing machine's memory dump is known as memory forensics. It is performed to explore and recognize attacks which normally do not leave behind any detectable tracks on hard drive data. Critical data regarding attacks exists only in system memory. This includes information about network connections, account credentials, chat messages, encryption keys, running processes, injected code portions, and Internet usage history. Every program including malicious or benign is loaded in the memory. Due to this memory forensics is of prime importance (Lord 2018). Mining of forensics data from hard disk, USB devices, CD, DVD, flash drives, floppy disks is called as disk forensics. Analysis of collected digital evidences includes image analysis, keywords, cookies, temporary files, time line scrutiny, mailbox analysis, recovery of deleted items, registry investigation, database examination, partition study, format analysis, and data slicing and its investigation (Prem et al. 2017). Evidence collection and analysis is elaborated in succeeding section.

7.8 Evidence Collection and Analysis

Proper evidence collection is very necessary for the cybercrime analysis. Due care should be taken while collecting available evidence either in physical or digital form. Cybercrime analysis consists of extensive data which poses the secure storage problem. Forensics investigators have to play with this stored data for long durations on trial and error basis. Figure 7.3 illustrates various steps involved in the evidence collection mechanism.

The traces obtained by investigators involve an Intrusion Detection System and firewall logs, logs generated by network services and applications, packet captures by sniffers. During data transactions, extensive data is generated by the network traffic which becomes critical for the people involved to analyze it and get the clue (Khobragade and Malik 2014). Data acquisition mechanism for cyber forensics is explained in subsequent section.

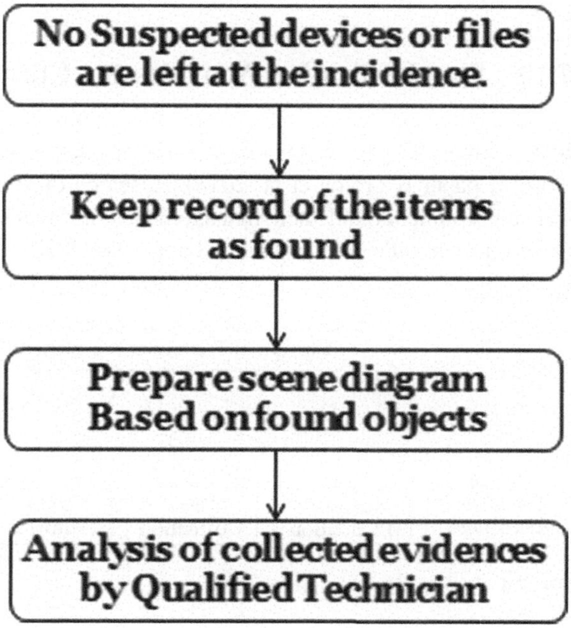

Fig. 7.3 Evidence collection guidelines

7.9 Data Acquisition

The forensics investigator has to obtain information conventionally from switched off device. He/she have to copy the whole data bit by bit carefully. The investigator has to work on these data files and reach towards the culprit. Recently, this mechanism is transformed into the real time data transfer system from turned on devices to his/her storage device. For online data transfer, the assessor has to run a little program on the suspect device which can acquire data to examiner's storage device. The essential steps involved in the forensic examination are depicted in Fig. 7.4.

Professional forensic examiners must be properly trained; their testing devices and software must be up to date; they should have knowledge about the legislative rules and regulations, and they should always be well equipped with the data acquisition toolkit, etc. Forensic evaluation contains instruction reception, risk analysis, and law enforcement decisions. Onsite data collection is performed while doing the Forensic investigation. It involves data mining, sorting, and labeling, etc. Data Analysis on the collected raw data is a very crucial part. After that, the investigator has to create an analytical report based on their findings. While writing the report, the non-technical language should be used which makes the report reader friendly. This data has to be presented in different meetings or teleconferences. Last but not the least is the time for taking reviews. This review is from right or bad experiences, findings and its correctness, timeliness and ultimately it is all about learning from previous mistakes (Baier 2011).

7.10 Standardization Activities for Client Side Analysis

State of the art cyber forensics analysis of client side is discussed in the research work of paper (Coronel et al. 2018). Research on digital evidences gathered from web environments and their management with analysis is presented with suggestions on standards improvement. Papers (ISO/IEC 27037 2012; ISO/IEC 27042

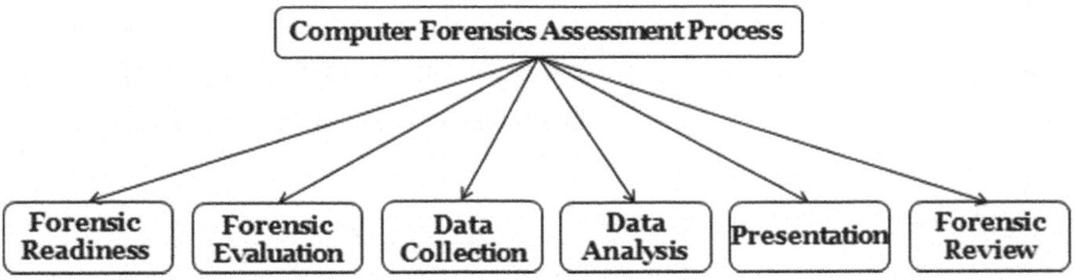

Fig. 7.4 Forensic investigation process

2015; ISO/IEC 27041 2015; ISO/IEC 27017 2015; ISO/IEC 27050 2017) provide information about standards useful for forensic environments are analyzed that includes

(1) ISO/IEC 27037—Review guidelines for identification collection acquisition and preservation of digital evidence.
(2) ISO/IEC 27042—Guidelines for the analysis and interpretation of digital evidence.
(3) ISOIEC 27041—Guidance on suitability and adequacy of incident investigative method assurance.
(4) ISO/IEC 27017—Code of practice for information security controls specifically for cloud services.
(5) ISO/IEC 27050—Provides requirements and guidance on activities in electronic discovery.

7.11 Summary

A lot of attention is needed towards the research is Cyber Forensics so that the cybercrime cases can be resolved in short time spans. The devices and the techniques to be employed should be very smart. When the ever growing information and communication technologies go hand in hand with cyber forensics, then we may witness robust security against cyber-crimes.

References

Baier H (2011/2012) Data acquisition and foundations of file system analysis. Hochschule Darmstadt, CASED, WS 2011/2012

Choi Y, Lee J-Y, Choi S, Kim J-H, Kim I (2016) Introduction to a network forensics system for cyber incidents analysis. In: 18th international conference on advanced communication technology (ICACT)

Coronel BD, Cedillo P, Campos K, Camacho J (2018) A systematic literature review in cyber forensics: current trends from the client perspective. IEEE Third Ecuador Technical Chapters Meeting (ETCM), pp 1–6

ISO/IEC 27017 (2015) Information technology—security techniques—code of practice for information security controls based on ISO/IEC 27002 for cloud services

ISO/IEC 27037 (2012) Information technology—security techniques—guidelines for identification, collection, acquisition and preservation of digital evidence. https://www.iso.org/standard/44381.html

ISO/IEC 27042 (2015) Information technology—security techniques—guidelines for the analysis and interpretation of digital evidence. https://www.iso.org/standard/44406.html

ISO/IEC 27041 (2015) Information technology—security techniques—guidance on assuring suitability and adequacy of incident investigative method

ISO/IEC 27050-3 (2017) Information technology—security techniques—electronic discovery—Part 3: code of practice for electronic discovery. https://www.iso.org/standard/66231.html

Kent K, Chevalier S, Grance T, Dang H (2006) Guide to integrating forensic techniques into incident response-recommendations of the National Institute of Standards and Technology, U. S. Department of Commerce

Khobragade PK, Malik LG (2014) Data generation and analysis for digital forensic application using data mining. In: Fourth international conference on communication systems and network technologies (CSNT), pp 458–462

Lord N (2018) What are memory forensics?—a definition of memory forensics. Digital Guardian Blog on Data Protection. https://digitalguardian.com/blog/what-are-memory-forensics-definition-memory-forensics

Paganini P (2012) Analysis of cybercrime and its impact on private and military sectors. PenTest Mag Audit Stand 03

Prem T, Paul Selwin V, Mohan AK (2017) Disk memory forensics—analysis of memory forensics frameworks flow. In: International conference on innovations in power and advanced computing technologies

Raftopoulos E, Dimitropoulos X (2013) Understanding network forensics analysis in an operational environment. IEEE Security and Privacy Workshop (SPW), pp 111–118 (2013)

Saibharath S, Geethakumari G (2015) Cloud forensics: evidence collection and preliminary analysis. In: IEEE international advance computing conference (IACC)

Scarfone K, Grance T, Masone K (2008) Computer security incident handling guide. National Institute of Standards and Technology, Mar 2008

Sekgwathe V, Talib M (2012) Cyber forensics: computer security and incident response. Int J New Comput Archit Their Appl 2(1):127–137

Threat Encyclopedia (2013) 7 cybercrime scenarios you should avoid. TrendLabs Security Gallery

West-Brown MJ, Stikvoort D, Kossakowski KP, Killcrece G, Ruefle R, Zajicek M (2003) Handbook for computer security incident response teams (CSIRTs), Apr 2003

Chapter 8
Cloud Computing

Cloud computing is a platform that provides state of the art information technology resources which are cheaper, scalable and easy to manage. But the cloud properties such as high scalability, quick deployment, dynamic resource range, high computing power, high bandwidth, etc. are exploited by cyber criminals as the tools for introducing new cybercrimes. These cybercrimes include DDoS as a service, Botnet as a service, Malware as a service, password cracking, BotClouds, C&C servers, Warez as a service, etc. To combat these cyber-attacks, the ICT system needs to be more pro-active. Fault tolerance in cloud computing platforms and applications is a crucial issue. This question is particularly challenging since cloud computing relies by nature on a complex splitting into many layers. This chapter describes various fault tolerance techniques with their pros and cons with future research directions.

8.1 Introduction

NIST defines cloud computing as, "Cloud Computing is a model for enabling convenient, on-demand network access to a shared pool of configurable computing resources (e.g., networks, servers, storage, applications, and services) that can be rapidly provisioned and released with minimal management effort or service provider interaction."

Another definition of cloud computing comes this way, it is a type of parallel and distributed system which consists of a group of interconnected and virtualized computing machines that are presented as a single computing and storage resource and has dynamic provisioning capability. Through the service level agreements and negotiation in between service provider and consumer, it can be made available to the users (Buyya et al. 2009).

Generally, cloud service provider has massive data centers spread over almost all geographical area to provide cloud services to the users. e.g., AWS-Amazon web

R. Prasad and V. Rohokale, *Cyber Security: The Lifeline of Information and Communication Technology*, Springer Series in Wireless Technology,
https://doi.org/10.1007/978-3-030-31703-4_8

Services and Azure: Microsoft cloud platform, etc. effectively. Based on pay per use strategy, this kind of data center can provide massive services with virtual machine.

Different layers in the Cloud Computing are depicted in Fig. 8.1. These layers include infrastructure containing computing devices, back storage system and networking components, platform counterpart includes database management, identity verification system, runtime service provisions, and queue management, application layer in cloud computing takes care of variety of applications provided by cloud computing such as collaboration, communication, finance services, content monitoring, etc.

The chapter is organized as follows. Threats to cloud computing environment are discussed in Sect. 8.2. Section 8.3 discusses cloud computing architecture. An overview of cyber attacks on cloud computing are elaborated in Sect. 8.4. Section 8.5 throws light on the fault tolerance technique. Fault Tolerance mechanisms are explained in Sect. 8.6. Section 8.7 illustrates state of the art security solutions available for cloud computing. Last Sect. 8.8 summarizes the chapter.

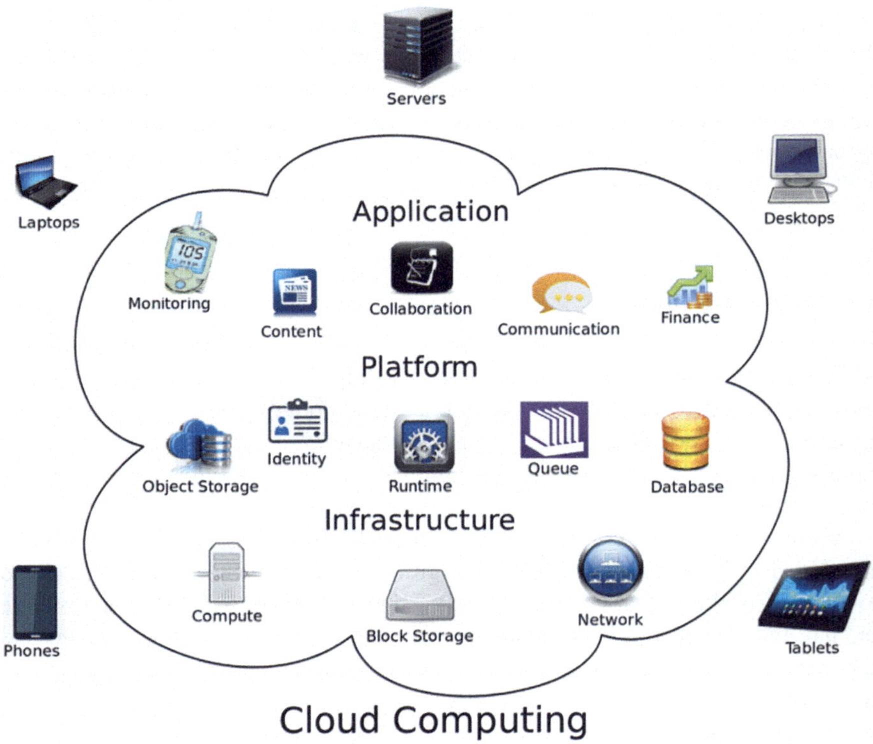

Fig. 8.1 Cloud computing model

8.2 Threats to Cloud Computing

Though so many processes carefully engineer the complete underlying infrastructure, the data centers are subject to a high number of failures because of great complexity of system and increasing the number of users acquiring the cloud-based services, the commodity components that expose the hardware to go on the condition that it was not actually designed for (Feller et al. 2012). Table 8.1 enlists the data breaches and records compromised since 2005. It is of prime importance to find the faults in the cloud computing systems and checks the fault tolerance of the system so that it can help in building new cyber security mechanisms.

The failure directly affects as the reduction in overall reliability of the cloud computing service, because of which fault tolerance becomes very much important for users as well as service providers. At the time of unexpected failures, the non-stop and accurate service provision becomes possible due to it. Traditional ways of achieving fault tolerance requires the users to have complete knowledge of the underlying mechanism but the actual scenario is that the scope of the user is only up to the users application and due to the failure of the data center the system may fail, which is entirely outside the scope of user's organization, as well as due to the pay-as-per-use bushiness model of cloud computing the service providers will never provide the system architectural details to the users.

In general, failure is nothing but system deviates from fulfilling the intended functionality or expected behavior, a failure happens due to error: that means the system is reaching to some invalid state. Fault tolerance is the ability of the network

Table 8.1 Data breaches and records compromised (Lord 2017; Kuranda 2016)

Sr. no.	Data breaches	Records compromised (in millions)	Year
1	Privacy rights clearinghouse	816	2005
2	TK/TJ Maxx	94	2007
3	Heartland's payment processing system	130	2008, 2009
4	Sony PlayStation network	77	2010
5	Sony online entertainment	24.6	2011
6	Evernote	50	2013
7	Living social	50	2013
8	Target	70	2013
9	Ebay	145	2014
10	Verizon data breach	700	2014
11	Home depot	56	2014
12	JP morgan chase	76	2014
13	Anthem	80	2015
14	MySpace users	360	2016
15	21st century oncology	2.2	2016

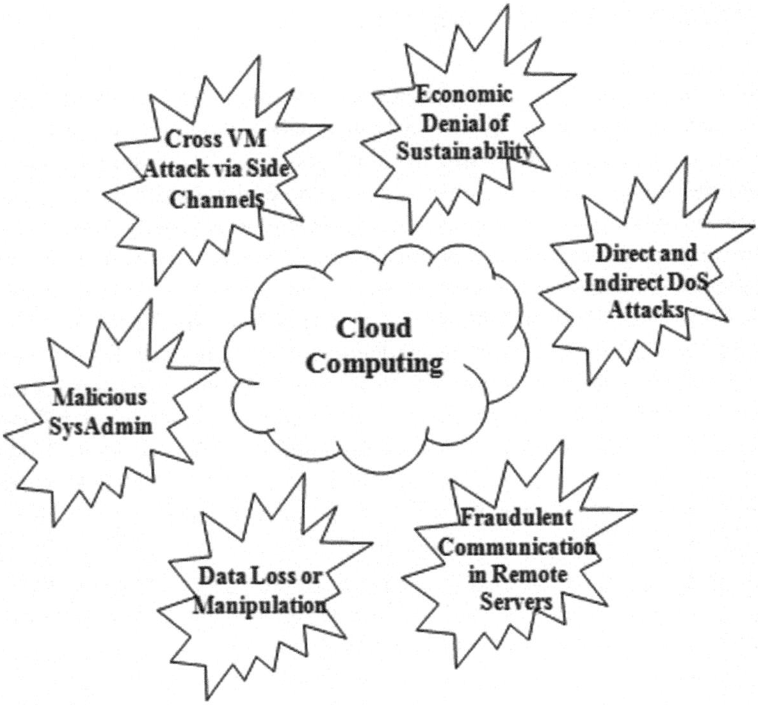

Fig. 8.2 Security threats to cloud computing

to perform its adequate function even the failures are present. The objective of this chapter is to develop the understanding of fault, different behavior of faults that may appear in cloud computing infrastructure, what are the different fault tolerance mechanisms and how these errors can be handled to improve the reliability and availability of the system in cloud computing environment.

There are numerous security threats to cloud computing environment as depicted in Fig. 8.2. Data loss and manipulation, malicious system administration, cross Virtual Machine (VM) attack through side channels, economic denial of sustainability, direct and indirect DoS attacks, deceitful communication in remote servers are some of the security threats posing as challenges in cloud computing. More common security threat issues like authentication, authorization, integrity, trust, and non-repudiation are also there in cloud computing infrastructure and services.

8.3 Cloud Computing Architecture

Cloud computing architecture has four different layers as illustrated in Fig. 8.3. According to the IETF (Internet Engineering Task Force) providers of Infrastructure as a service offers computer—physical or more often virtual machines and other

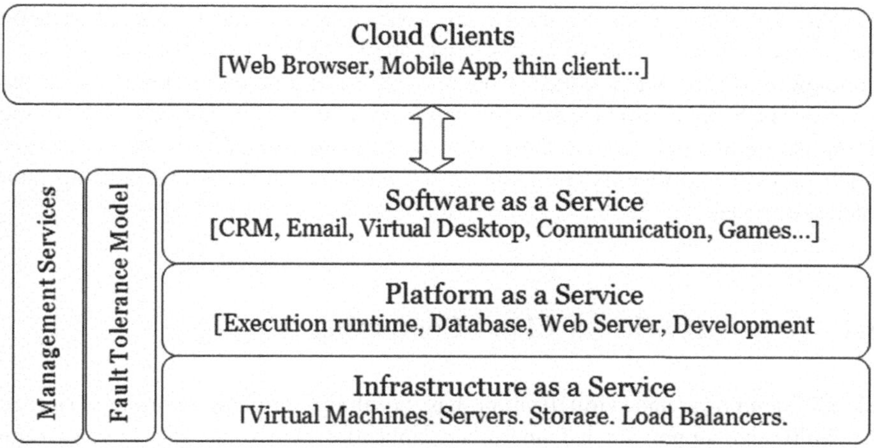

Fig. 8.3 Layered architecture of cloud computing

resources. (A hypervisor such as Xen, Oracle Virtual Box, KVM, VMware, ESX, ESXi or Hiper-V runs the virtual machines as guests). Infrastructure as a service clouds also offers additional resources such as a virtual machine disk image library, raw block storage, load balancers, IP addresses, Virtual local area networks (VLANs) and software bundles. Infrastructure as a service cloud provides these resources on demand for their large pools installed in data centers.

Cloud computing has three development models such as private cloud, public cloud, and hybrid cloud. Private cloud is cloud infrastructure operated solely for a single organization. When the services are rendered over a network that is open for public use is called the public cloud. Hybrid cloud is a composition of two or more clouds offering the benefits of multiple deploying models.

There are four basic layers in the cloud infrastructure like infrastructure as a service (IaaS), platform as a service (PaaS), software as a service (SaaS), and cloud clients. IaaS and PaaS mostly include the physical cloud resources with core middleware functionalities. Application services are provided by SaaS including services provided to lower layers. These kind of services are normally developed and provided by third party service providers.

PaaS model contains computing platform that may include an operating system, programming language, and web browser. Using this platform as a model service client can develop their application and run their software solutions on a cloud without buying and managing underlying hardware and software layers. Some platform as service providers are Microsoft Azure and Google App Engine, etc.

The SaaS model cloud service providers usually install and operate application software in the cloud and the users can access that software from the cloud clients. Cloud users are nothing to manage about the cloud infrastructure and the platform. This reduces the burden of the user to install and execute the application on the users system, which also simplifies the maintenance and support.

Typically, if the failure occurred at any layer of this architecture that impacts on the services offered by above layers, for example, failure took place at user-level Middleware (PaaS) may produce errors in SaaS and if a failure occurred the virtual machine (IaaS), may produce errors in above both layers. The failure in IaaS or the hardware significantly affects the system as a whole. Because of this, it is very much important to identify the faults in the physical hardware and develop fault tolerance model.

8.4 Cyber Attacks on Cloud Computing

Cloud computing and virtualization has no doubt brought enormous storage capability and support for IoT and other applications but it has also increased the attack space and possibilities. Nine various attacks are categorized based on the factors like cloud infrastructure or network attack surface, victim entity, side channels, etc. Free of charge usage of cloud services and stealing of cloud resources without paying for it comes under theft of service attack. Denial of service attack has four subclasses such as distributed denial of service (DDoS), Http based DDoS, Xml based DDoS, REST based DDoS, and light traffic DoS termed as shrew attack. Under this attack, the services are made unavailable to the user, for getting unauthorized access to the critical information, malicious code is wrapped inside the Xml signatures, and sensitive information is obtained though browser history via unsecure Http browser. Leakage of credential information, user data and abnormal behavior of cloud machine is observed under cloud malware injection attack. Cross virtual machine side channel attack is of two types. First one is about disturbing or changing the timing of side channels and energy consuming side channels. Outflow of user data or cloud infrastructure related sensitive information is leaked through this attack (Khalil et al. 2014).

Targeted shared memory attacks generally include cloud resource information leakage, user private information outflow, and also it provides support for side channel attacks and malware injection attacks to enter into the cloud system. Phishing attacks try to get unauthorized access to the user's private information by installing a malicious code into the target user's computing device, it also changes the behavior of cloud computing system from normal to abnormal, and also it makes the server unavailable for the customers.

Botnets are the stepping stone attacks used to obtain the access to the cloud resources in unauthorized way. All these known attacks are illustrated in Fig. 8.4. Stealing of user data and sensitive information ate the main intents behind this kind of attacks. Illegal access to user data and deletion of user data comes under audio Steganography attack. It also refrain the cloud services from being available to the end users. VM rollback attack initiates the brute force attack in the cloud services thereby trying to damage cloud infrastructure. Main aim behind this attack is the leakage of sensitive information and user credentials (Khalil et al. 2014).

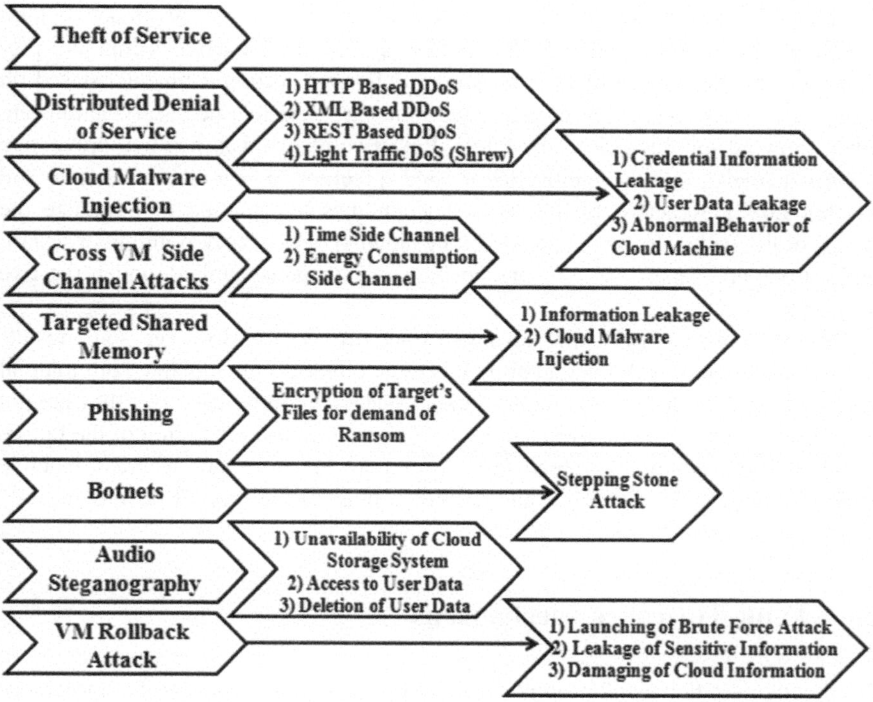

Fig. 8.4 Possible attacks on cloud computing

8.5 Fault Tolerance—An Overview

In a large scale distributed cloud computing environment, various types of faults may attack and cause the failure of the system. The failures that occur in cloud computing can be classified into two classes as data failures and computation failures. Data failures occur due to the corrupted data contents, missing source data, etc. and computational failure occurs due to all types of hardware issues like infrastructure failures, faulty VM's, storage issues, etc., (Li et al. 2010). According to Microsoft, almost 80% of the failures and crashes are possible to eliminate if top 20% of them are fixed in time. This is called as Pareto Principle or 80/20 rule (Zheng et al. 2010).

Based on the statistical information obtained from large-scale studies on data center failure behavior performed using the extensive collection of servers (approximately 100,000 servers) discuss that each server in data center contains multiple processors, storage disks, memory modules, network interfaces, etc. and the annual failure rate is around 8%, and the repairing costs the amount to 2.5 million dollars approximately spend for 100,000 servers (Vishwanath and Nagappan 2010).

Fault tolerance policy has basically two main types including proactive fault tolerance policy and reactive fault tolerance policy. In proactive technique, precautions are taken ahead of time by taking defensive accomplishments. Based on historical studies and analysis, underlying faults are detected, necessary actions are taken to avoid failures due to the faults. Proactive mechanism has two preventive actions namely software rejuvenation and preemptive migration (Ganga and Karthik 2013). Reactive fault tolerance techniques are applied to minimize the effects of the faults already happened in the cloud system. Check pointing or restart, replication, and snapshot are some of the mechanisms applied under reactive techniques.

Both the policies have some advantages and drawbacks. However, some results show that proactive fault tolerant system is more efficient than sensible fault tolerant policy. Even though dynamic fault-tolerant systems are more effective, they are not often used compare to reactive techniques. This is because, prediction of the failure is bit difficult and on that prediction system is less affected, whereas reactive methods are easy to implement and are not applied at the time of development.

8.6 Fault Tolerance Mechanisms

For the highly reliable and promptly service providing cloud system, fault tolerance mechanism plays a vital role. Various researchers working on the fault tolerance techniques for cloud computing have come up with innovative techniques. Fault tolerance mechanism can be divided into three classes like check pointing techniques, data replication techniques and snapshot mechanisms as shown in Fig. 8.5.

- **Check-pointing Fault Tolerance Mechanism**

Here if failure occurs on the go, then the system can turn back to the latest checkpoint and restart from there. As compared with the large-scale distributed computing environments the cloud computing environment is highly complex and of virtualized nature. It poses two problems including how often we can insert checkpoint and can save on the go systems. If check-points are inserted too

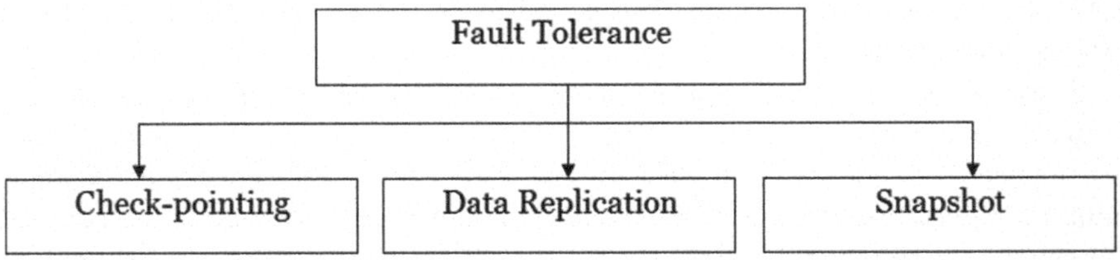

Fig. 8.5 Classification of fault tolerance mechanisms

frequently, the overhead of large check-pointing such as large storage will be required due to a large number of inserted checkpoints which is quite doubtful in the cloud environment. Large operating costs are to be paid if the checkpoints are inserted more frequently with a large number of operations whenever the failure occurs. Usually the question is that which check-pointing strategy should be used for cloud computing environment? There are three check-pointing strategies as comprehensive check-pointing strategy, incremental check-pointing strategy, and hybrid check-pointing strategy (Li and Lan 2011; Marzouk and Jmaiel 2011).

Figure 8.6 shows the check-pointing strategies in which, the cross symbol indicates the occurrence of a failure, the timestamp indicates the periodic check-point taken at the particular time, dark timestamp indicates full checkpoint and standard (white) timestamp indicates incremental checkpoint.

a. **Total check-pointing strategy**: The advantage of the scheme is it reduces the re-computing time of the system whenever the system rolls back from failure. Practically, the comprehensive check-pointing strategy is not that much suitable for cloud computing environment as it requires a significant check-pointing overhead for saving the whole system running states periodically to the storage system.
b. **Incremental check-pointing strategy**: In the additional check-pointing procedure, the first checkpoint contains complete system running state, subsequent checkpoints contains only pages that have been modified since the previous checkpoint. The advantage of incremental checkpoint strategy is that it reduces the check-pointing storage overhead by saving the pages changed that have been modified instead of keeping the whole system running state.
c. **Hybrid check-pointing strategy**: Hybrid check-pointing strategy is the combination of both clear check-pointing strategy and incremental check-pointing strategy. In the case of failure, the system will restart from the state of the last full check-pointing (Li and Lan 2011).

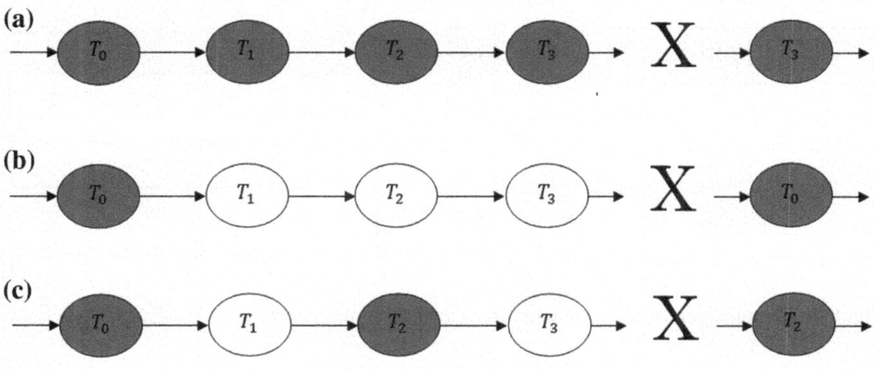

Fig. 8.6 Check-pointing strategies. **a** Full check-pointing strategy, **b** incremental check-pointing strategy, **c** hybrid check-pointing strategy

The check-pointing mechanism gained significant attention over the past few years, in the context of fault tolerance research. The classification of the check-pointing mechanism is full check-pointing and additional check-pointing mechanism depending on whether the whole system running states or newly modified page states are saved. The check-pointing mechanism can further be classified as the local check-pointing mechanism and global check-pointing mechanism based on whether the check-pointing data is collected, locally or globally. The check-pointing mechanism can also be classified as disk check-pointing mechanism and diskless check-pointing mechanism, based on saving the check-pointing data on disk or not as shown in Fig. 8.7.

- **Data Replication Fault Tolerance Mechanism**

Data imitation is the mechanism used for improving the data availability as well as offering the fault tolerance of the users by providing various replicas with a coherent state of the same service (Lei et al. 2008; Yuan et al. 2010). Data replication can be classified into two groups from replication strategy as static data replication and dynamic data replication mechanisms. In static replication process the replication strategy is predefined; in contrast, active replication creates and deletes the replicas as the access pattern changes. In research paper (Shvachko et al. 2010), a static cloud data replication mechanism is proposed, which specifies the number of replicas for each file, block size and replication factor. The state of the art comparative Table 8.2 helps to identify the benefits and limitations of various techniques which are discussed here and earlier in previous sections.

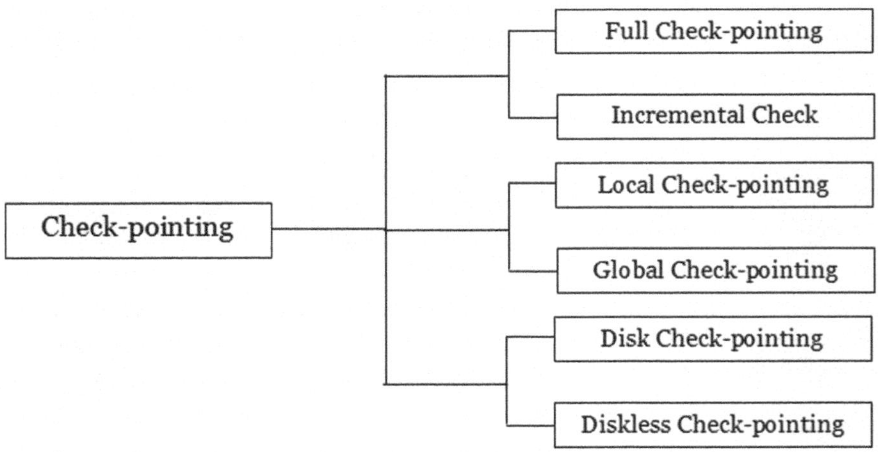

Fig. 8.7 Classification of check-pointing fault tolerance mechanism

Table 8.2 Comparative state of the art for fault tolerance in cloud computing

Sr. no	FT model	Proactive	Reactive	FT technique	Programming framework	Fault detected	Performance	Response time	Reliability
1	Low latency fault tolerance (LLFT)	No	Yes	Replication, check-pointing	C++, Linux	Application failures	High	Average	High
2	Assure	Yes	No	Check-point, retry, self-healing	JAVA	Host, network failure	High	Average	Average
3	Fault tolerance manager (FTM)	No	Yes	Replication and check-pointing	–	Application / network failures	Average	Average	High
4	SHelp	No	Yes	Check-pointing, self-healing	SQL, HTTP, Linux	Application failures	Average	High	Low
5	Self-tuning failure detector (SFD)	Yes	No	Check-pointing, retry	–	Host, network failure	High	Average	Average
6	Byzantine fault tolerant framework BFTCloud	No	Yes	Replication	–	Application failures	High	Average	High
7	Virtualization and fault tolerance (VFT)	Yes	Yes	Replication	Java, HTML	Application failures	High	High	High

8.7 State of the Art in Cloud Security

Cloud federation formation process with security perspective is presented in paper (Halabi et al. 2018). Cloud infrastructure security risk level assessment is performed aiming to find the reference security risk level for the cloud service providers. Fractional hedonic coalitional game is proposed in this research work for modeling the federation formation process. Based on the reputations of the CSP's, the cooperation strategy is decided. Objectionable CSP's are avoided in the cooperation game strategy based on pre-designed evaluation criteria for security risk prevention and alleviation.

The research work in paper (Puri and Agnihotri 2017) presents the cloud security strategy starting from attack data collection to design and implementation of honeypots for cyber attack mitigation. Ubuntu based OwnCloud intrusion detection system with honeypot implementation is carried out here. Collection of the real statistical data, analysis of the collected data and automated classification is done by using signature based engines with new aspects. This system is capable of detecting the well-known and unfamiliar attacks aiming cloud infrastructure.

If the user's sensitive information stored with parent business entity is needed to be shared with partner organizations, then the user's identity needs to be verified alongwith his/her access rights. This kind of information transaction poses serious security and privacy issues. In the cloud environment, three standard identity protocols Security Assertion Markup Language (SAML), Open Authentication (OAuth), and OpenID Connect (OIDC) are popularly known for this purpose. The research work in (Naik and Jenkins 2016) puts forth a working prototype and significant investigation of these three open standard protocols. This paper also presents the assessment measures necessary for analysis task. The potential strengths and weaknesses of these three protocols are discussed for all types of cloud computing models. Out of all three protocols, OpenID Connect is found to be the best suited protocol for almost all the cloud computing requirements.

Threat model helps the security personnel in knowing and developing a defense strategy for combating against the possible attacks and vulnerabilities. It is basically the necessary requirement for security optimization. According to (Cable 2018), the security operations teams get advantages from creation of threat model for cloud computing infrastructure. In the process of development of threat model, there are different approaches to be taken care of include operationalization, hardening, and security automation. Five steps for creation of a threat model are given as follows.

- Get the appropriately educated team for the purpose
- Model the threat system
- Take into account risk mitigation
- Build a strong plan and allocate responsible persons for the risk
- Constant monitoring of cloud infrastructure

Due to shared network environment of cloud computing, it has to face lot of security and privacy issues including authentication and authorization. Risk Adaptable Access Control (RAdAC) framework is presented in (Abdullah and Bakar 2018) which addresses security issues in cloud computing. Two factor authentication mechanism for tight security is proposed with the RAdAC framework in this paper. For dynamic cloud environment, two tier security provisions prove to be more beneficial in terms of security and privacy provision.

Security attacks on cloud computing and promising solutions such as XML signature wrapping attacks, browser security and vendor lock in attacks and evaluation of security solutions for them are presented in the paper (Alshammari et al. 2017). Secure coding is applied as a security solution against XML signature wrapping attacks. XML signature is used as a defense mechanism for browser security attacks. Model driven architecture is used to combat for lock-in attacks.

An educational cloud is the famous cloud computing technology used for sharing of education resources in online education systems (Nie et al. 2018). Data loss is the most severe security issue in such kind of educational cloud environments. Load runner and AppScan tools are used for the security analysis of educational clouds. These tools help in protecting against various security and privacy issues faced by these educational cloud platforms.

8.8 Summary

Cloud computing is a booming field and very important for upcoming networks like WSN, IoT, ITS, etc. Many research experts are working on developing a standalone foolproof security system to tackle multiple types of faults, attacks and failures. Because of its virtualized structure, cloud environment is open to variety of attacks. To provide security and privacy are the major challenge. Secure and private cloud computing system is the urgent need of time. Huge research scope is there for the budding researchers in this field.

References

Abdullah S, Bakar KAA (2018) Security and privacy challenges in cloud computing. In: Cyber resilience conference (CRC)

Alshammari A, Alhaidari S, Alharbi A, Zohdy M (2017) Security threats and challenges in cloud computing. In: IEEE 4th international conference on cyber security and cloud computing

Buyya R, Yeo CS, Venugopal S, Broberg J, Brandic I (2009) Cloud computing and emerging IT platforms: vision, hype, and reality for delivering computing as the 5th utility. Future Gener Comput Syst 25(6):599–616

Cable P (2018) How to create a threat model for cloud infrastructure security. Threat Stack AWS SECURITY Blog. https://www.threatstack.com/blog/how-to-create-a-threat-model-for-cloud-infrastructure-security

Feller E, Rilling L, Morin C (2012) Snoze: a scalable and automatic virtual machine management framework for private clouds. In: Proceedings of CCGrid'12, Ottawa, Canada, pp 482–489 (2012)

Ganga K, Karthik S (2013) A fault tolerant approach in scientific workflow systems based on cloud computing. In: IEEE international conference on pattern recognition, informatics and medical engineering (PRIME), pp 387–390

Halabi T, Bellaiche M, Abusitta A (2018) A cooperative game for online cloud federation formation based on security risk assessment. In: 5th IEEE international conference on cyber security and cloud computing (CSCloud) and 4th IEEE international conference on edge computing and scalable cloud (EdgeCom), pp 83–88

Khalil IM, Khreishah A, Azeem M (2014) Cloud computing security: a survey. Open Access J Comput 3:1–35

Kuranda S (2016) The 10 biggest data breaches of 2016 (So Far). CRN article, 28 July 2016. http://www.crn.com/slide-shows/security/300081491/the-10-biggest-data-breaches-of-2016-so-far.htm

Lei M, Vrbsky SV, Hong X (2008) An on-line replication strategy to increase availability in data grids. Future Gener Comput Syst 24(2):85–98

Li Y, Lan ZFREM (2011) a fast restart mechanism for general checkpoint/restart. IEEE Trans Comput 60(5):639–652

Li J, Humphrey M, Cheah YW, Ryu Y, Agarwal D, Jackson K, van Ingen C (2010) Fault tolerance and scaling in e-Science cloud applications: observations from the continuing development of MODISAzure. In: 6th International Conference on e-Science (e-Science), IEEE, pp 246–253

Lord N (2017) The history of data breaches. Digital Guardian Blog, 27 Jan 2017. https://digitalguardian.com/blog/history-data-breaches

Marzouk S, Jmaiel M (2011) A survey on software check-pointing and mobility techniques in distributed systems. Concurr Comput 23(11):1196–1212

Naik N, Jenkins P (2016) An analysis of open standard identity protocols in cloud computing security paradigm. In: IEEE 14th International conference on dependable, autonomic and secure computing, 14th international conference on pervasive intelligence and computing, with 2nd international conference on big data intelligence and computing and cyber science and technology congress

Nie W, Xiao X, Wu Z, Wu Y, Shen F, Luo X (2018) The research of information security for the education cloud platform based on AppScan technology. In: 5th IEEE international conference on cyber security and cloud computing (CSCloud) and 4th IEEE international conference on edge computing and scalable cloud (EdgeCom)

Puri S, Agnihotri M (2017) A proactive approach for cyber attack mitigation in cloud network. In: International conference on energy, communication, data analytics and soft computing (ICECDS)

Shvachko K, Kuang H, Radia S, Chansler R (2010) The Hadoop distributed file system. In: Proceedings of 2010 IEEE 26th symposium on mass storage systems and technologies (MSST 2010). IEEE Press, New York, pp 1–10

Vishwanath K, Nagappan N (2010) Characterizing cloud computing hardware reliability. In: Proceeding of SoCC'10, Indianpolis, IN, USA, pp 193–204

Yuan D, Yang Y, Liu X, Chen J (2010) A data placement strategy in scientific cloud workflows. Future Gener Comput Syst 26(8):1200–1214

Zheng Z, Zhou TC, Lyu MR, King I (2010) FTCloud: a component ranking framework for fault tolerant cloud applications. In IEEE 21st international symposium on software reliability engineering (ISSRE 2010), pp 398–407, Nov 2010

Chapter 9
Internet of Things (IoT) and Machine to Machine (M2M) Communication

Kevin Ashton, form Auto-ID center at MIT first coined the term Internet of things in 1999. IoT is nothing but the convergence of wired technologies, wireless communications, micro-electromechanical devices, micro-services software systems building small single function modules, and Internet connectivity. IoT is basically a network of connected physical objects and they communicate with each other through Internet. Variety of sensors are used for sensing the surrounding situations and RFIDs are used for providing IP addresses that is network identities to these objects. Without any human intervention, these things can communicate with objects and bring easiness into human lives.

In recent years, with the development of wireless sensor networks, breakthroughs have been made in Internet of Things (IoT) with wide applications. As any serious contribution for advance Internet of Things must necessarily be the result of synergetic activities conducted for various different fields of knowledge, such as telecommunications, informatics, electronics and social science. Machine to machine communication (M2M) is considered to be the subset and integral part of IoT. M2M makes use of wireless as well as wired networks for the communication or relaying purpose.

9.1 Introduction to IoT and M2M

IoT mainly consists of three components like things, people and systems as illustrated in Fig. 9.1. Things include variety of machines, devices, sensors, computer products, vehicles, etc. People include workers, consumers, employees, partners and customers. Systems is a vast concept containing business applications, analytics systems, enterprise resource planning (ERP), product lifecycle management (PLM), customer relationship management (CRM), data warehouses, control systems and various other necessary functionalities.

R. Prasad and V. Rohokale, *Cyber Security: The Lifeline of Information and Communication Technology*, Springer Series in Wireless Technology,
https://doi.org/10.1007/978-3-030-31703-4_9

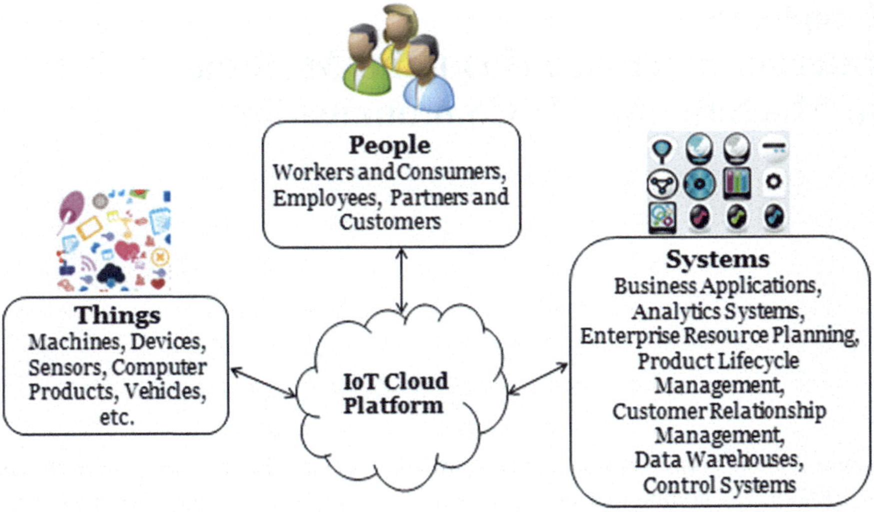

Fig. 9.1 Internet of Things (IoT) conceptual view

Internet wave during 1960s stirred the world and changed the whole aspects of human life. Today Internet has penetrated very deep into the human life and society including rural sectors of the developing countries. It was the biggest technological revolution due to which the globe is connected through the computer networks. Because of this, Internet is known as network of networks. Now is the time of IoT revolution wherein billions of connected objects will be interacting with every other object wherever located in the world. The IoT devices such as sensors will generate huge amount of data. The storage, processing, and presentation of this data in an efficient form is a big challenge (Nagesh 2013; Atzori et al. 2010).

Almost every aspect of human life is covered by machine to machine communication. Industrial wave of progress has succeeded because of the strong base of M2M only. It is used largely in power sector, transportation segment, industrial control, retail zone, public service management, water, oil, gas, and energy industries among many. M2M is found suitable for many applications like safety monitoring, automation industry, vehicle theft protection, auto sales, mechanical maintenance, public transport management, and many other smart city solutions in combination with IoT. It can transmit data through cable, wireless channel, mobile communication, or other means (Du and Chao 2010). M2M communication has to face lot of threats including denial of service attack, device triggering attack, access priority attack, external interface attack, false network attack, tamper attack, SMS spoofing attack, etc. (Hussain et al. 2016). Table 9.1 illustrates the clear differences in between M2M and IoT (Fig. 9.2).

The organization of the chapter is as follows. Section 9.2 gives information regarding some of the use cases of M2M and IoT. Security issues in M2M communication are discussed in Sect. 9.3. Section 9.4 describes state of the art in

Table 9.1 Difference between M2M and IoT

Sr. no.	Machine to machine (M2M) communication	Internet of Things (IoT)
1	M2M is a technology that allows machines to communicate or relay information as per need over some protocol such as internet protocol (IP)	IoT is interaction of objects surrounding us. Objects without intelligence or without connected to Internet can also be brought into the IoT via smart phone acting as gateway to the Internet
2	Relies on point to point communication using embedded hardware module and a cellular telecommunication or wired network	Relies on IP based networks to interface device data to cloud or middleware platform
3	Hardware reliance makes M2M not capable to handle big data	Software reliance makes IoT capable to handle big data with ease
4	Only device data	Integration of device and sensor data
5	E.g., Communication with machines such as refrigerator, washing machine, smart meters through IP over wireless or wired	E.g., Interaction with newspaper advertisement via short code or movie poster via near field communication (NFC)
6	Limited scalability due to requirement of incremental hardwired connections and SIM card installation so M2M is referred as plumbing	Scalable, Universal Enabler and makes use of different standards

Fig. 9.2 M2M communication conceptual view

M2M security. Standardization activities related to M2M security are conversed in Sect. 9.5. Section 9.6 gives idea about security and privacy issues in IoT. State of the art in IoT security is discussed in Sect. 9.7. Section 9.8 explains standardization activities for IoT security. Last Sect. 9.9 summarizes the chapter.

9.2 Use Cases of M2M and IoT

Millions of machines communicating with each other and billions of smart objects are interacting with all other interconnected objects worldwide is going to be the scenario by 2020. From not connected and isolated objects to the interconnected smart objects, the journey of M2M and IoT evolution is interesting. M2M is a subset but the integral part of IoT. These approximately 50 billion of smart objects while interacting with each other are going to give rise to enormous amount of data. Handling of this data with keeping privacy and security intact is the next big challenge for the growth of M2M and IoT. From isolated, monitored, remotely controlled and managed machines in automation industry to intelligence embedded smart systems like smart homes, offices, connected cars, intelligent transport system, smart grid, etc., and then IoT is the stunning journey.

9.2.1 M2M Use Cases

CISCO VNI survey in 2014 says that the number of connected machines is going to reach around 7.3 Million by 2018. The M2M adoption growth is variable across different territories of the world. In the developing countries, the growth is seen to be high for automotive and consumer electronics market. Commercial market focus is mainly to better understand customers' needs, ecosystem needs, to improve service delivery, reduce support costs and topping the market (Intelligence 2014). Now a days, M2M is adopted in alms every aspect of human life. Some of the use cases include everyday human activities like photo uploading from still camera, text or content download to eBook reader device, remote control of home and office appliances such as fans, lights, air conditioners, uploading of surveillance data, inventory management, etc., (ETSI TR 102 857 2013). These use cases are illustrated in Table 9.2. The common blocks of the M2M system include M2M device, gateway, wireless network such as cellular or IP network, M2M service platform with management server and intended M2M application.

Table 9.2 Use Cases of M2M and IoT

Sr. no.	M2M use cases	IoT use cases
1	Photo uploading from still camera 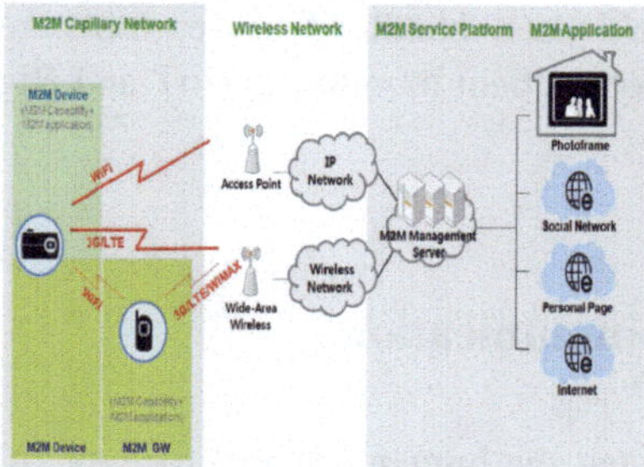 Source: Use Cases of M2M applications for Connected Consumer, ETSI TR 102 857 V1.1.1, Aug 2013	Retail and logistics 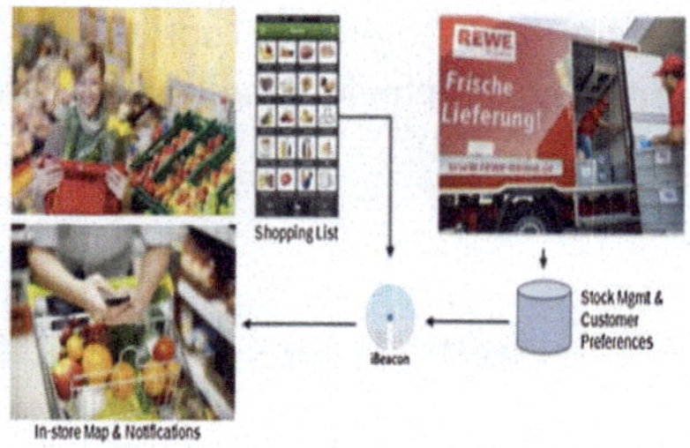 Source: IoT and Big data Use Cases, BOSCH blog.bosch-si.com/…/ internet-of-things-iot-and-big-data, Aug 2014
2	Content download to eBook reader device 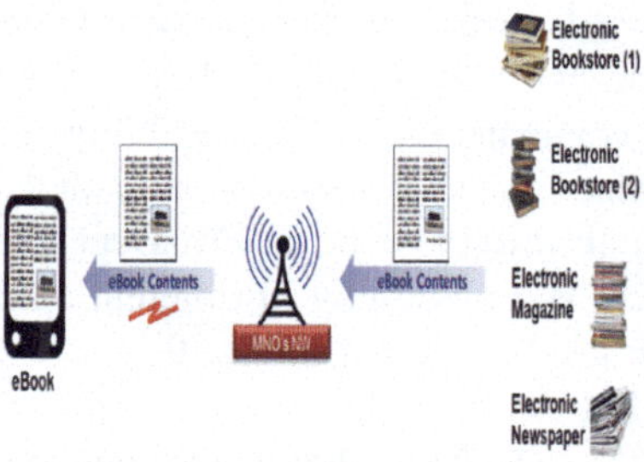 Source: Use Cases of M2M applications for connected consumer, ETSI TR 102 857 V1.1.1, Aug 2013	Systematic capturing of field data  Source: IoT and Big Data Use Cases, BOSCH blog.bosch-si.com/…/ internet-of-things-iot-and-big-data, Aug 2014
3	Remote control of home appliances	Handheld power tools
4	Uploading of surveillance data	Remote vehicle condition monitoring
5	Inventory management	Remote patient monitoring

9.2.2 IoT Use Cases

IoT is considered as an extension of M2M technology. M2M is the connection of devices and data collection from them. IoT is the interconnection of all such devices for specific use cases and decision making from the data obtained. The growth of IoT interconnected objects is increasing enormously.

According to the research work in Olavsrud (2015), IoT enabled devices need three main capabilities like awareness, autonomous and actionable. These devices should provide context aware services means they should sense the surrounding environment and depending on these inputs, they should autonomously act and these devices are needed to be capable enough to command and control the situation and they should be able to take the decisions if necessary. There are numerous use cases of IoT including transportation companies making use of IoT data for reduction in fuel consumption, transition from conventional street lights to LED street lights which don't need regular maintenance but can report if there is need of repairing (Fleisch 2010). Table 9.2 lists some of the use cases of IoT like retail and logistics, capturing of field data, handheld power tools, remote monitoring of vehicle conditions and remote patient monitoring (IoT and Big data Use Cases 2018).

9.3 Security Issues in M2M Communication

Machine type communication has penetrated the human life very deeply. Almost every object around human beings everyday is somehow connected to M2M that may be through wired or wireless connectivity (Gazis et al. 2012). The ever increasing usage of connected devices in the industrial plants and in the supply chains gives rise to vulnerabilities. The attack surface vicinity is very large. The connected devices are always listening with their ports open and they do not know who is talking to them and who is trying to command them. The scalability of M2M is huge and due to which denial of service attack is very common threat that ultimately result into the network congestion (Stojmenovic 2013). Depending on various parts of M2M communication, there is a possibility of different attacks at device level, service level or infrastructure level as shown in Fig. 9.3.

As depicted in the figure, attacks on M2M can be at device level, services level or infrastructure level. At device level, software can be used to reprogram the victim device with malign intents, the sensitive information can be extracted with the help of component analysis, or the hardware or software pieces can be replaced. False traffic can be injected into the M2M network at service level, or confidentiality can be broken through traffic analysis, and component operations can be modified with bad intention. At infrastructure level, subscriber information can be extracted from the control system, or network components and services can be identified and

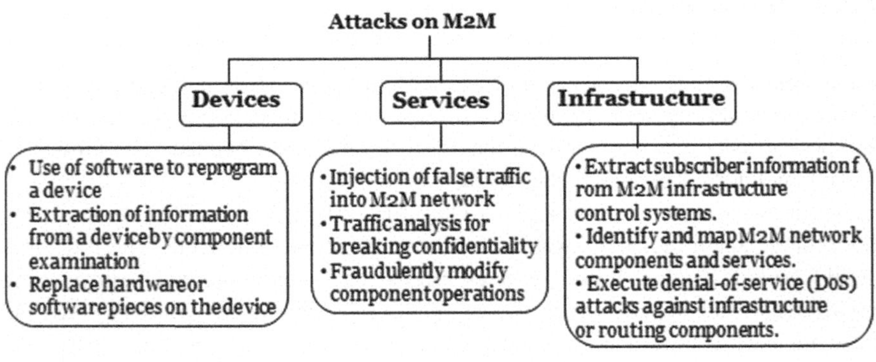

Fig. 9.3 Various possible attacks on M2M

planned, and denial of service attack can be introduced by compromising the network devices.

Man-in-the-middle attack (MITM) can capture the data, alter the data, and can transmit the modified data through the network resulting in the loss to the user. Usually, the attackers get entrance in the system, they can alter the insightful information, they can delete the important data, or they can repeat the user transactions (Ghidini et al. 2014). Security threats are widely broadened over all the protocol layers of the M2M network. Once the security provision at physical layer and data link layer is appropriate, then the network layer and application layer can also be straightforwardly protected. For designing security solutions such as intrusion detection, key management, trust management, authentication, privacy protection, cross layer approach is found more suitable. Layer wise threat scenario for M2M is depicted in Fig. 9.4.

Physical layer attacks include active and passive attacks. Passive attacks normally perform traffic analysis by eavesdropping the wireless communication among the network entities. Active attacks are more dangerous and are responsible for network traffic jams and targeted small duration attacks like scrambling. Collision, overhearing, idle listening, retransmission, and cross-examination are some of the actions performed under exhaustion type of attack. MAC layer attacks include Sybil attack in which once the attacker gets access to the network, it takes multiple identities and origins more harm to the MAC layer with its malign functionalities. In de-synchronization attack, the network nodes are made to de-synchronize and detach from the network. False network attacks, tampering and SMS spoofing come under the network layer attacks. Application layer attacks comprise of replay attack, snooping, session hijacking, and false network attacks (Hussain et al. 2016).

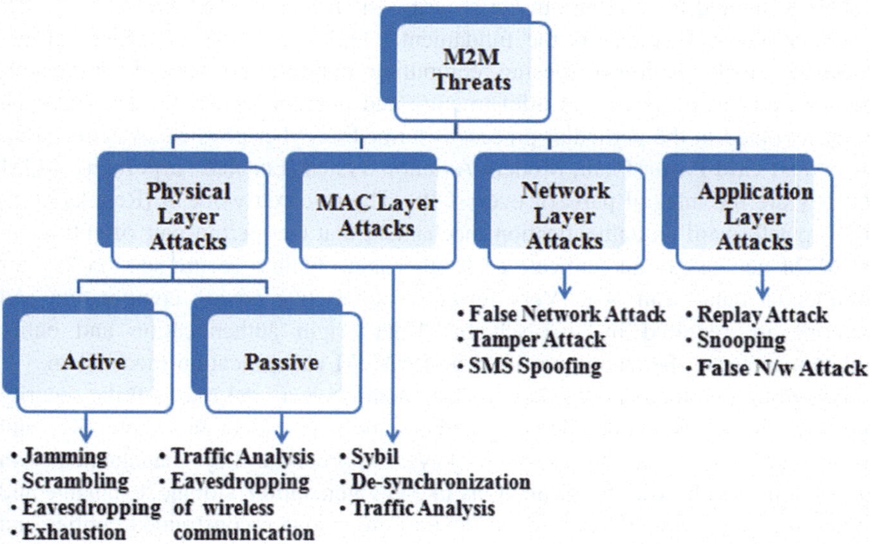

Fig. 9.4 Layer wise attack scenario in M2M

9.3.1 Practical Attacks Scenarios of M2M

As the world is progressing from M2M to IoT, variety of new vulnerabilities are found which are attacking M2M in different ways. These attacks either take place through modified software codes or the compromised hardware devices. These scenarios include tracking devices like GPS and GPRS, stealing of cars through duplication of electronic keys, compromising smart meters, hacking of Insulin pumps and heart rate monitors, etc. Cyber security which can combat such vulnerabilities is highly awaited. With the growth of M2M entities in Millions, researchers have tough challenge to develop robust security algorithms which can withstand such attacks. Some of the known attacks include Russians break through the US power grid from 2016 to 2018, Ukraine Blackout is another example of attack happened in 2015 and 2016, Marriott data breach is the famous case, social media attacks are making waves, Stuxnet was the worst attack happened on the M2M in 2010, Soviet gas pipeline explosion was the largest non-nuclear explosion happened due to attack on M2M. Smart meters, autonomous vehicles, smart homes, smart offices, health monitoring devices are the M2M attack targets.

9.4 M2M Security-State of the Art

For information security, intrusion detection with more accurate degree is essential. Efficient intrusion detection system for the cyber physical system with M2M and IoT is presented in paper (Belenko et al. 2018) that makes use of artificial Neural Networks for correct intrusion detection. Generative Adversarial Neural Network (GAN) is applied for finding out the security glitches and cyber threats.

Cyber physical system is the fundamental building blocks of critical infrastructures. Amalgamation of sensing, computing, control, and networking capabilities into physical objects and infrastructure and interconnection to web and each entity involved in the particular process is termed as cyber physical system (CPS). Eg., Smart Grid Power Plant, Modern Aviation System Enabled Airport, etc. M2M and IoT are the integral parts of every CPS. The research work in (Renuka et al. 2019) put forward an authentication mechanism that can permit pair of entities in the M2M network to authenticate each other and obtain a secure session key for confidential data swap over. Very fewer resources like power, computation and memory are required by this scheme. Data origin authentication and entity authentication are the techniques used under M2M authentication mechanism.

For group communications, key management plays a vital role. All the security building blocks like authentication, authorization, confidentiality, integrity, and privacy are based on the usage of keys. Appropriate key management can successfully handle variety of functions like key generation, storage, exchange and updation of the required keys. M2M devices are resource constrained entities and because of this public key cryptography is not fitting them. So, private key cryptosystem proves to be a better solution for M2M. Some of the researchers have come up with symmetric key solutions between M2M device and remote server, some have proposed Elliptic Curve Cryptography (ECC) based public/private key agreement protocol, Identity based cryptography (IBC) is put forward for key management, key management in between gateway and M2M device is designed based on ECC and certificate agent, Diffie Hellman key exchange algorithm is used for key management in between device and gateway for M2M key management.

M2M provides lot of benefits to the user and industry community but it needs to protect user privacy which is related to sensitive personal data or localization information. Solutions towards safeguarding the user privacy consists of IBC and pseudonyms based privacy protecting scheme when charging electric vehicles in the smart grid system, location privacy solutions are proposed for pay as you drive and electronic toll systems, Walsh codes and secret keys are used for privacy protection against the data collected by smart meters, etc. Confidentiality issues are tackled with the help of symmetric key cryptography which is less resource consuming for M2M scenario. Some solutions have come up with NTRUEncrypt, a lattice based cryptosystem, a multivariate cryptosystem, and usage of an optimized family of prime numbers to provide confidentiality for M2M communication. For integrity assurance, schemes such as bandwidth efficient cooperative authentication have been designed by the researchers (Barki et al. 2016).

9.5 M2M Security Standardization Activities

Standardization plays a vital role in developing a single market policy. Standards provide safety, reliability, support of government policies and legislation, interoperability, business benefits such as market access, scale economies and awareness about consumer choice (ITU-T Focus Group 2014). Standardization activities related to M2M security mainly include variants of 3GPP and ETSI as mentioned below:

- 3GPP SA3 MTC which takes care of mobile communication related security portion.
- 3GPP SIMTC it assures security enhancement for machine type communication (MTC).
- ETSI TISPAN is related to M2M threat analysis and privacy facet of M2M.
- ETSI TC SCP and M2M takes care of M2M outline aspects.
- ETSI TC M2M is regarding authentication, integrity, and confidentiality for M2M gateway to infrastructure crossing point.

9.6 Security and Privacy Issues of IoT

Number of connected devices is increasing by a massive number with the IoT. With it, the number of possible victims to cyber attacks is also growing on large scale. IoT has evolved with a great pace but the security shield is in the limited form till now. IoT is nothing but connecting the devices through Internet. With everytime, everything, and everywhere connectivity, comes security and privacy threat. The most general cyber attacks on IoT include botnets, man in the middle attack, data and identity Theft, Social engineering and denial of service attack (GlobalSign Blog 2016).

Botnet is the network of infected bots. For employing magnificent scale attacks, botnets are used.

IoT devices under botnet are termed as thingbots which can include all connected things, like computers, laptops, smartphones, tablets, and other personal electronic gadgets. Botnets are applied by attackers for stealing sensitive private information, exploiting online banking data, and employing DDoS attacks through spam and phishing emails. Man in the middle attack is taken place when eavesdropper tries to intercept the communication between two entities. Attacker can modify the message in transit without knowledge of the sender and receiver. Because of the provision of the limited security, the connected things are the easy targets for the attackers.

Data transactions from the internet connected devices should be safeguarded, otherwise attackers can misuse it and user has chances to lose precious data and money. Information about user's personal identity can be easily gathered from social media, personal digital gadgets and various apps mounted on the smartphone. User's personal identity can be stolen and attacker poses as if genuine identity and performs various malign activities.

Social engineering is the tactic of influence users so that attacker can easily take out private information from them. Users are decoyed to expose their passwords or financial information. These attacks are performed with the help of phishing emails posing as if legitimate financial organization or online shopping site. Usually when the services are unavailable, denial of service attack takes place. If it is a distributed denial of service attack infecting group of computing devices like botnet, then the whole cyber physical system like power plant, aviation system, train service can come to standstill. Instead of money or information stealing intent, in DoS and DDoS attacks, the loss of reputation for the organization is the biggest loss (Whitter-Jones 2018).

9.6.1 IoT Security Threats and Challenges

Digital Era brings major potential threats to user privacy, user's personal sensitive data such as pictures, videos, messages, banking credential information, and many other things that are linked to IoT. These IoT threats can be used as weapons by the hackers for cyber war. Everything connected to Internet is prone to security threat. By 2025, almost 75 Billion IoT devices will come into existence. Because of the resource constrained nature of IoT devices, they cannot employ complex security protocols which require large memory space and big computation capability. And due to this, IoT devices are more vulnerable to security threats. Some of the probable security threats and challenges are shown in Fig. 9.5.

IoT device manufacturing and supplying industries should provide service updates on timely basis. Otherwise these connected IoT devices and customers may face potential cyber attacks and data breaches. If the IoT devices are hijacked, then they can be used to perform any malign action such as sending spam and/or phishing emails. Compromised IoT devices can be used to act as Botnet nodes and may be used to introduce DDoS attack in the massive IoT network. Without encrypted communication messages among IoT devices, cloud and IoT devices, gateways and IoT devices, possibility of man-in-the-middle attack increases. Users should more often change the password of their computing machines and electronic gadgets. Use of default passwords for long time is more risky and prone to security attacks like brute force attack targeting passwords. Remote access to the IoT

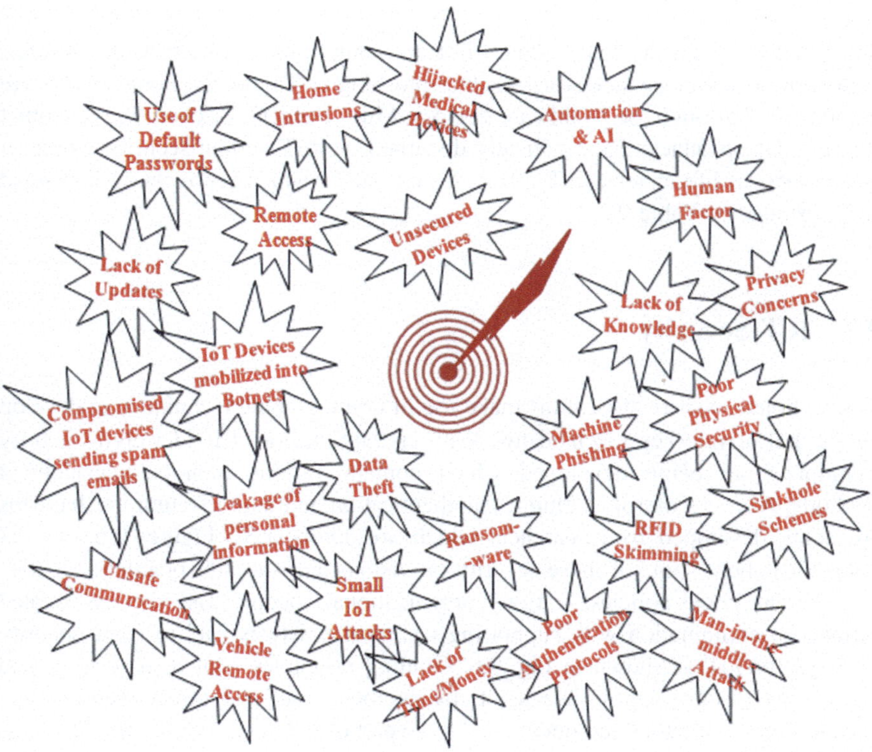

Fig. 9.5 IoT security threats and challenges

devices such as home appliances or office cameras can land them into attack troubles through weak communication links. By gaining IP addresses through unsecured IoT devices, hackers can get idea about user's location, their residential address and their routine pattern of staying away from home. This can lead to home intrusions and thefts.

Remote access to the vehicles in the Intelligent Transportation System (ITS) can be a high risk of vehicle hijacking and may lead to vehicle theft, accidents or life threats. Attackers lock user's smart devices through ransomware attack and demand digital money like bitcoin to unlock them. Data theft from corporate sector may lead to loss of customer information such as their names, addresses, credit card numbers, financial credentials, etc. Hijacking of healthcare IoT devices may target killing of the patient. If hackers get access to automation and artificial intelligence platforms, it can result in massive destruction and huge loss of assets. Machine phishing may lead to standstill condition of the manufacturing plant. Poor physical security, weak authentication protocols, pathetic safeguarding of user privacy, MITM attack, sinkhole attack, and RFID skimming are the big challenges for IoT (Hasan 2019).

9.6.2 Practical Attack Scenarios of IoT

Smart objects from the IoT internetwork are compromised to attack the systems with various malicious techniques such as phishing or Thingbots. Some attacks are implemented through the healthcare devices. Mirai Botnets, Brickerbot, the Botnet Barrage, DDoS attacks have critically disturbed the IoT system security. Some of these practical life attacks (IoT 2017; Wallen 2017) which are recently happened are illustrated in Table 9.3.

9.7 IoT Security

Conventional security algorithms make use of cryptographic techniques to develop public and private key mechanisms. With encryption, lot of light weight security techniques are being developed which contain various Identity management methods, Access Control techniques, authentication and authorization mechanisms are being developed by researchers which are appreciated. Figures 9.6 and 9.7 gives idea about some of these security provisions and superior practices for IoT.

Different good policies, various organizational, people and process related actions, in combination with supporting technical practices lead to basic security dealings for IoT as shown in Fig. 9.6. Guiding principles related to security and privacy by design, appropriate asset management, risk and threat identification ahead of time with good judgment need to be part of IoT security policies. Dealings related to organizations, people, and process must contain the end of life support for things with proven solutions, provision of good management of security

Table 9.3 Latest IoT attack incidences

Sr. no.	IoT attack	Description	Year
1	Mirai Botnet	Largest DDoS attack affected the functionality of Twitter, the Guardian, Netflix, Reddit, and CAN	2016
2	Hacking of cardiac pacemaker devices from St. Jude	Depletion of pacemaker batteries, and/or modifying pacing and shock rates	2017
3	Owlet WiFi baby monitor vulnerabilities	Resulted into heart troubles in babies	2016
4	TRENDnet webcam hack	Faulty software affected various services from home security to baby monitoring	2012
5	The Jeep hack	Using vehicle CAN bus protocol, total control of the SUV Jeep was taken and it was misguided	2015
6	Stuxnet	Computer worm targeting SCADA systems and is responsible for damaging Iranian Nuclear Plant	2010
7	Cold in Finland	DDoS attack shutting down heater system in two buildings of Finland	2016
8	Brickerbot	Almost 60,000 Indian modems and routers of BSNL and MTNL were hacked leading to loss of Internet connectivity	2017
9	The Botnet Barrage	Unknown university 5000 IoT devices including vending machines were hacked resulting in slow or lost connectivity	2017

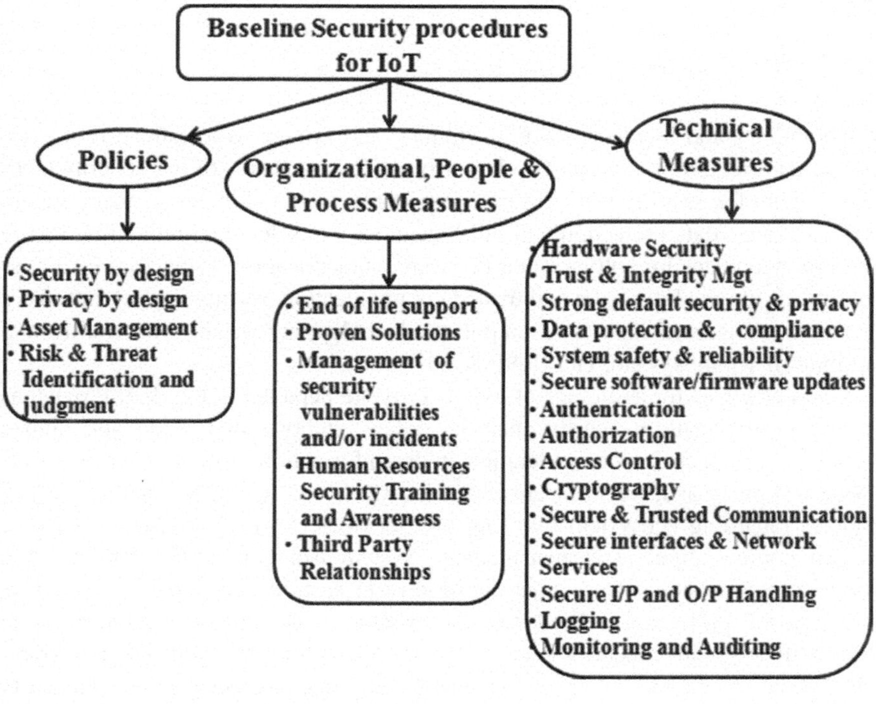

Fig. 9.6 IoT fundamental security procedures

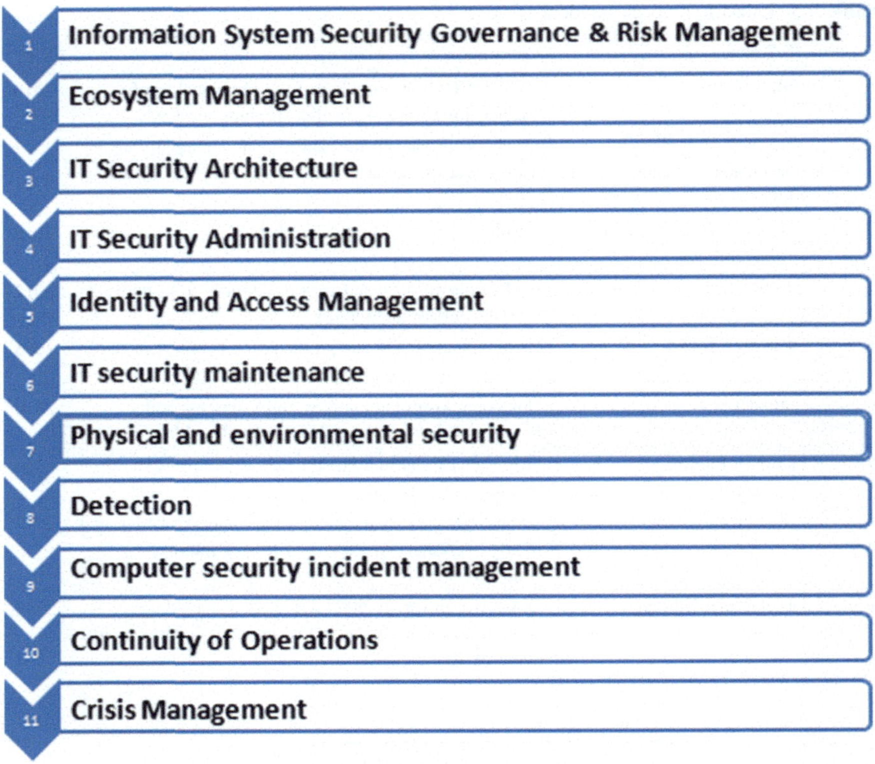

Fig. 9.7 IoT security dealings and good practices

vulnerabilities and incidents, security training and awareness to human resources, and policies regarding third party relationships. Technical side of IoT security takes into account the security of hardware components, trust and integrity management, privacy issues, data protection and fulfillment, system safety and reliability, secure and timely updates for software and firmware, authentication, authorization, access control, cryptography, secure and trusted communication, secure network interfaces and services, secure input and output conduct, logging details, monitoring and auditing of whole system, etc. (ENISA 2017).

Good practices for assurance of IoT security are depicted in Fig. 9.7. Provisions related to information security analysis, policy, authorization, audit and human resources are taken into consideration under information system security governance and risk management mechanism. Ecosystem management contains security policies regarding ecosystem mapping and relations. Security dealings related to system configuration, asset management, system isolation, traffic filtering, and cryptography are considered under the head of IT security architecture. IT security management protection related to administration accounts and administration information systems. Security issues regarding IT security maintenance procedures and remote access are separately considered. Physical and environmental security takes care of the security of physical infrastructure and environmental safety against pollution and other factors. Safety measures related to detection, logging, log correction, and analysis come under detection mechanism. Computer security and incident handling system takes into account information security incident analysis, response, and report preparation. Business continuity management and disaster recovery management are considered under stability of operation. Crisis management organization and process related security issues are taken into consideration by separate entity (ENISA 2017).

9.8 Standardization Activities for IoT Security

With the world marching towards fifth generation mobile communication, the IoT interconnected objects are predicted to reach the figure of around 50 Billion. Standardization of upcoming IoT use cases is a great challenge for security

Table 9.4 Standardization activities related to IoT and its security, privacy

Sr. no.	Standardization efforts aimed for	Standardization body	Provision details
1	Overall IoT network stack	ITU-SG 20	IoT, smart cities, and communities, IoT applications, identification, security, privacy, and openness
		IEEE P2413	IoT architectural framework with applications, networking, data communication, and sensing Addresses security and authentication issues
		3GPP NarrowBand IoT (NB-IoT)	Support IoT technology for low power wide area network (LPWAN), for indoor coverage with low cost and long battery life massive IoT deployment. Addresses security
		IETF-LPWAN IPv6 over LPWAN Working Group	Lower layer offerings on low power wide area network from SigFox, LORA Alliance, 3GPP, etc. Addresses security
		OASIS-IBM-MQTT and MQTT-S	Provides security standards, privacy yet to be considered
		oneM2M	Presents specifications for architecture, APIs, security, and interoperability guidelines and certification for IoT/M2M devices and applications. Basic security architecture provision
		W3C-WoT	Reduction of IoT fragmentation, reduction of development cost
2	Application specific standardization efforts	Fairhair Alliance driven by Philips, Siemens	Technologies for lighting control and building automation
		THREAD Group driven by Google	Home automation and smart home solutions

researchers. IoT related standardization activities are listed in Table 9.4 which mainly includes standards related to overall IoT Network stack and application specific standardization efforts (Pal et al. 2018).

9.9 Summary

M2M and IoT are the key technologies which are going to play major roles in the next generation ICT and the due attention must be given towards the cyber security related issues concerned with M2M and IoT. Extensive research is needed in this area specifically considering computational limitations of small size and limited computational memory of the sensors involved in these networks.

References

Atzori L, Iera A, Morabito G (2010) The Internet of Things: a survey. Elsevier, Amsterdam

Barki A, Bouabdallah A, Gharout S, Traoré J (2016) M2M security: challenges and solutions. IEEE Commun Surveys Tutor 18(2)

Belenko V, Chernenko V, Kalinin M, Krundyshev V (2018) Evaluation of GAN applicability for intrusion detection in self-organizing networks of cyber physical systems. In: International Russian automation conference (RusAutoCon)

Du J, Chao SW (2010) A study of information security for M2M of IOT. In: 3rd international conference on advanced computer theory and engineering

ENISA (2017) Baseline security recommendations for IoT in the context of critical information infrastructures, Nov 2017

ETSI TR 102 857 (2013) V1.1.1, use cases of M2M applications for connected consumer, Aug 2013

Fleisch E (2010) What is the Internet of Things?—an economic perspective. Auto-ID Labs White Paper WP-BIZAPP-053, Jan 2010

Gazis V, Sasloglou K, Frangiadakis N, Kikiras P (2012) Wireless sensor networking, automation technologies and machine to machine developments on the path to the Internet of Things. In: 16th Panhellenic conference on informatics

Ghidini G, Emmons SP, Kamangar FA, Smith JO (2014) Advancing M2M communications management: a cloud-based system for cellular traffic analysis. 15th international IEEE symposium on a world of wireless, mobile and multimedia networks (WoWMoM)

GlobalSign Blog (2016) 5 common cyber attacks in the IoT—threat alert on a grand scale. https://www.globalsign.com/en/blog/five-common-cyber-attacks-in-the-iot/

Hasan M (2019) 25 most common IoT security threats in an increasingly connected world. UBUNTUPIT Open Source Trends. https://www.ubuntupit.com/25-most-common-iot-security-threats-in-an-increasingly-connected-world/

Hussain F, Ferdouse L, Anpalagan A, Karim L, Woungang I (2016) Security threats in M2M networks: a survey with case study. Int J Comput Syst Sci Eng

GSMA Intelligence (2014) Analysis from concept to delivery: the M2M market today, Feb 2014

IoT and Big data Use Cases (2018) BOSCH, Aug 2018. http://blog.boschsi.com/categories/manufacturing/2014/08/internet-of-things-iot-and-big-data-brought-together-in-commercial-use-cases

IoT for all Blog on "The 5 worst examples of IoT hacking and vulnerabilities in recorded history (2017) https://www.iotforall.com/5-worst-iot-hacking-vulnerabilities/

ITU-T Focus Group (2014) M2M service layer: Requirements and architectural framework. ITU-T Focus Group Technical Report on M2M Service Layer, Deliverable D2.1 "M2M service layer: Requirements and architectural framework"

Nagesh S (2013) Roll of data mining in cyber security. J Exclusive Manag Sci 2(5):2277–5684

Olavsrud T, CIO Blog on (2015) Internet of Things connections to quadruple by 2020, 20 Mar 2015

Pal A, Rath HK, Shailendra S, Bhattacharyya A (2018) IoT standardization: the road ahead. In: Sen J (ed) Internet of Things. IntechOpen Publishing

Renuka KM, Kumari S, Zhao D, Li L (2019) Design of a secure password-based authentication scheme for M2M networks in IoT enabled cyber-physical systems. In: IEEE access of special section on security and privacy for cloud and IoT

Stojmenovic I (2013) Large scale cyber-physical systems: distributed actuation, in-network processing and machine-to-machine communications. In: Mediterranean conference on embedded computing

Wallen J (2017) Five nightmarish attacks that show the risks of IoT security. Part of a Zdnet special feature: cybersecurity in an IoT and mobile world. https://www.zdnet.com/article/5-nightmarish-attacks-that-show-the-risks-of-iot-security/

Whitter-Jones J (2018) Security review on the Internet of Things. In: Third international conference on fog and mobile edge computing (FMEC)

Chapter 10
Smart Grid

Electrical power system developed few decades back is now being transformed into smart grid power systems due to the availability of many utilities like Smart metering system with auto monitoring quality of power, automatic control by Supervisory Control and Data Acquisition (SCADA) system, and usage of renewable or green energy sources for reduction of carbon footprint. Smart grid provides a solution for stability, reliability, efficiency, voltage regulation, cost, air pollution, and safety of energy systems. The smart grid system is designed for visualizing and monitoring of real-time demand of power and distribution of electricity as per load from consumers. Smart grid infrastructure system is a fusion of two infrastructures, i.e., physical and cyber infrastructure. It makes use of both the physical and digital infrastructure. Cyber infrastructure makes use of a communication network and SCADA system for control and trustworthy data communication in two way communication mode.

Smart Grid intelligently uses information technology supremacy to smartly distribute energy using two way communication protocol and at the same time, it is keeping the environment green by integrating renewable energy sources into the power grid. Communication technology is vulnerable to various security threats due to its nature. The key concepts to be taken into consideration while designing security for smart grid include confidentiality, integrity, availability and accountability.

10.1 Introduction to Smart Grid

Smart Grid infrastructure includes Advanced Metering Infrastructure (AMI), Demand Response (DR), Network Reliability (NR) Grid Optimization and Integration of Renewable Energy. AMI system collects all the data and information from loads of consumers and transmits it towards control center. AMI is configured as an infrastructure that integrates some of the technologies to reach its goals.

R. Prasad and V. Rohokale, *Cyber Security: The Lifeline of Information and Communication Technology*, Springer Series in Wireless Technology,
https://doi.org/10.1007/978-3-030-31703-4_10

Wireless mesh networks (WMNs) are mostly preferred as communication infrastructures based on IEEE standard, namely IEEE 802.11 s.

For security provision in Smart Grids, it is needed to develop a design model for risk identification and management. The design model will combine Secure Communication Protocol (SCP) designed for communication security and State Estimation Model (SEM) for secure verification, detection and prevention from attack and to provide protection for both physicals as well as cyber sides. The end to end access, security encryption model with secure key management can be integrated with protocol model to provide the solution for availability, privacy, Scalability and Denial of Service transmission between the smart meters and the other devices. Using this model, it is possible to monitor information from automated electric grid which can evaluate a solution for infrastructure problems such as power outages, zero faults, false data injection (FDI) in advanced metering infrastructures, end-to-end access protection, etc. Any error detected can be easily displayed and located to actual grid sections on the operators screen. The framework should also consider survivability to intrusions, customer service center problems, etc.

Address Resolution Protocol (ARP) is inefficient because of broadcasting of ARP request when creating and maintaining the ARP cache. Inefficiency degrades the performance, therefore for obtaining efficiency, researchers planned an efficient ARP proactive path request (PREQ) message as a part of the 802.11s standard. This specification uses media access control (MAC) address resolution for the period of routing tree formation and maintenance. Providing security by considering confidentiality, Integrity, availability and accountability is a big challenge.

The chapter is organized as follows. Section 10.2 elaborates smart grid network architecture. Cyber security for smart grid is discussed in Sect. 10.3. Section 10.4 throws light on advance metering infrastructure and smart metering systems. Risk management is described in Sect. 10.5. Section 10.6 gives idea about advance metering infrastructure security. AMI security techniques are explained in Sect. 10.6.3. Section 10.7 summarizes the chapter.

10.2 Smart Grid Network Architecture

Network Architecture of Smart Grids mainly consists of various Smart Grid Components involved and different layers used as Smart Grid Model Layers as discussed in following paragraphs.

10.2.1 Smart Grid Components

Electrical energy used in day to day life is generated from renewable and nonrenewable sources. Natural or nonrenewable sources available on earth are limited.

Therefore effective utilization of energy generated using nonrenewable sources is necessary. Power generated by using renewable sources such as thermal, wind, hydropower, tide, etc., is insufficient, and it is not cost effective. Research studies are directed towards finding an effective solution to renewable energy resources for generation of power. The initial power system was designed to transmit power from generation station to end user customer. The Flow of electricity was in one direction only, and central station controlled this flow. Due to recent advancements in technology now the unidirectional movement of power is transformed into the bidirectional flow. The electrical power system is now smart grid power systems due to availability like Advanced Metering Infrastructures (AMI), information and communication technology (ICT), Supervisory Control and Data Acquisition (SCADA) system. Smart grid is a bi-directional communication system that allows users to manage their energy service, as well as access smart grid convenience features, such as controlling power related operations, turning home power on/off reading current details of bills. To provide these functions, smart grid infrastructure is developed which consists of components, such as electrical energy physical structure, hardware, software and required communication network (Rahman et al. 2013). Components of smart grid system are shown in Fig. 10.1.

To meet the demand of consumer, smart grid infrastructure provides on-demand power supply. Smart grid controls total flow of monitoring according to needs of customer and increases efficiency of transmission from generation to consumer. The primary difference between smart grid and traditional power distribution system is

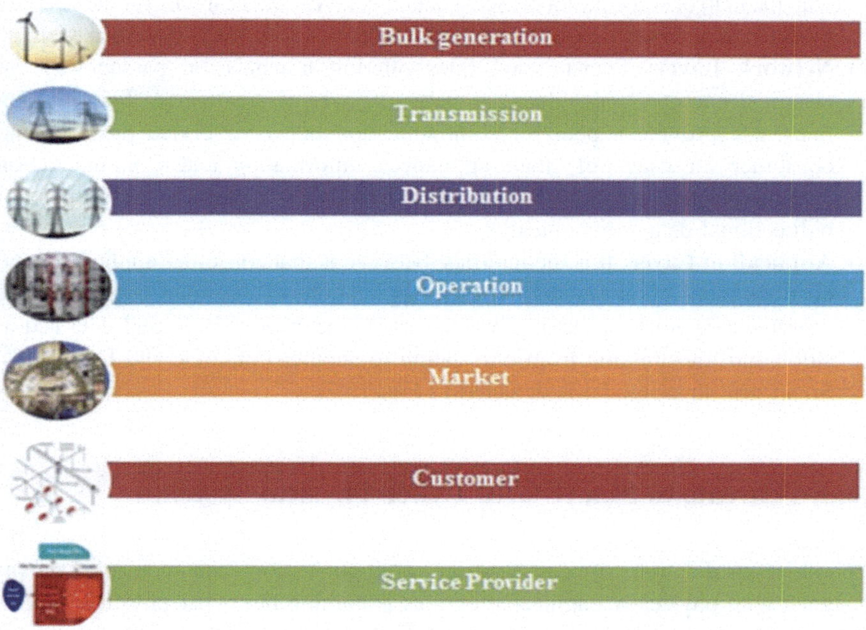

Fig. 10.1 Smart grid system components

that traditional power distribution network is the one-way transmission of power whereas smart grid system is bi-directional flow system and flow of power can be controlled by Supervisory Control and Data Acquisition (SCADA) system. By measuring the usage of the consumer, nature of user can be predicted, and smart grid systems can manage power.

10.2.2 Smart Grid Model Layers

Smart grid model design consists of mainly four layers including power generation plants considered as a physical layer, Supervisory Control and Data Acquisition (SCADA) system as data link layer, Power Transmission Lines as network layer, and Smart Grid Applications and Services as application layer.

(1) **Physical Layer**: Power generation physical infrastructure comes under physical layer consideration for smart grid. Renewable and nonrenewable sources generate energy at physical layer. Nonrenewable sources such as coal, natural gas, fossil fuels, etc. are limited hence efficient use of energy is necessary. Electric power is generated by rotation of turbine, rotation of these machine converts mechanical energy into electrical energy. Renewable energy sources include solar, wind turbines, tidal energy, vibration energy, etc. This generated power is transmitted through transmission lines.
(2) **Data Link Layer**: Transfer of data in between control center and consumer is done with the help of some protocols set at data link layer. Data Link Layer uses certain rules for Supervisory Control and Data Acquisition (SCADA) system as well.
(3) **Network Layer**: Transmission lines offering transmission passageway for electrical energy is considered as network layer of smart grid. Transmission lines also provide a path for communication between service provider and consumer for usage of smart appliances. Information and Communication Technology (ICT) makes use of duplex transmission lines for the purpose of bidirectional data communication.
(4) **Application Layer**: It is the interface between actual consumer application and host communication system. Different applications of the smart power grid are power management system, advanced smart metering system and data acquisition and transmission from one station to another (Vukovic and Dan 2013). Smart grid layer model is depicted in Fig. 10.2.

10.3 Threats to Smart Grid Cyber Physical System

Smooth consolidation of computation, communication and control systems designed with physically engineered entities is nothing but cyber physical system. Smart grid cyber physical system is prone to various threats because of presence of some weak counterparts in the computation, communication and control systems.

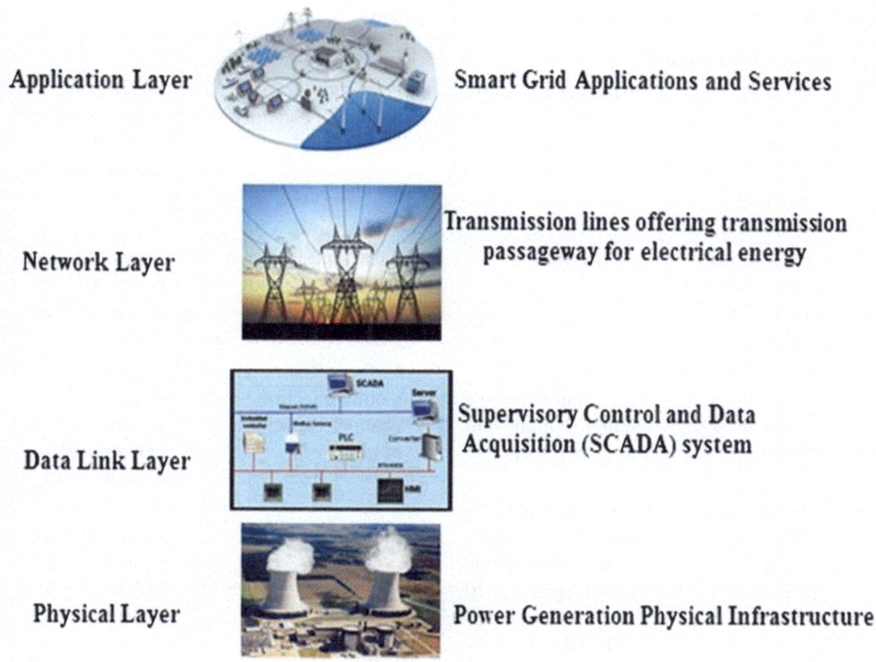

Fig. 10.2 Layer wise model of smart grid

Process control system or SCADA security has lot of threats including intrusion detection. Smart meter security threats contain IDS, meter data management, redundant smart meter reading. Smart meter data anonymization and privacy, smart system state estimation security has threats like false data injection attacks, communication channel flooding attack, etc. Smart grid communication protocol has lot of security threats such as problems in protocol design principles, real time smart meter communication threats and traditional cryptography security issues (Baumeister 2011). The threats to smart grid cyber physical system are shown in Table 10.1. Meter data management system in the advanced metering infrastructure has lot of serious threats like false data injection, data anonymization, etc.

Domain wise cyber security threats to smart grid cyber physical system are depicted in Fig. 10.3. Due to the distributive nature of smart green it is prone to natural threats due to weather and other general threats like cyber infrastructure malfunction, equipment crash, and cyber space failure. Bulk energy generation side threats include intimidation from non-conventional energy generation source like photovoltaic units. The attacks can hugely impact softwares involved, hardware equipments and communication channels. Denial of service attack on communication channels can harm the whole system. Control units can be remotely hijacked and controlled by malign actors. On the customer side, possible threats are the attacks on advanced metering infrastructure (AMI), hijacking on electric vehicles, service thefts and fraud, etc. Distribution domain risks contain the attacks on

Table 10.1 Threats to smart grid cyber physical system

Sr. no.	Security measures			
	Process Control System (PCS)/ SCADA security	Smart meter security	Power system state estimation security	Smart grid communication protocol security
1	PCS security risks (Jiaxi et al. 2006; Watts 2003)	IDS (Berthier et al. 2010)	False data injection attacks (Dan and Sandberg 2010; Kosut et al. 2010; Xie et al. 2010)	Protocol Design Principles (Khurana et al. 2010)
2	IDS (Valdes and Cheung 2009)	Redundant smart meter reading (Varodayan and Gao 2010)	Communication channel capacity (Li et al. 2010)	Real time communication (Zhang and Gunter 2010)
3	PCS security assessment techniques (Jiaxi et al. 2006)	Smart meter data anonymization and privacy (Efthymiou and Kalogridis 2010; Kalogridis et al. 2010; NIST 2010)		Smart meter communication (Bartoli et al. 2010; Li et al. 2010)
4				Cryptography (So et al. 2010)

physical distribution system, availability issues can be raised to refrain the competitor from smoothly supplying electricity supply to customers. Operations field faces authentication and authorization issues. Market domain has issues like physical hazards, availability issues, IP address spoofing network attack and so on. Service provider side has intimidations from DoS attack to traffic sniffing and threat to financial credentials. Transmission sector has threats from physical attacks and control system attack (Olmstead and Siraj 2011).

10.4 Attacks on Smart Grid

The smart grid system is a combination of hardware and software; the hardware is comprised of physical infrastructure for power transmission and communication networks to control operations. The communication system is the complex system which is an interconnection of the system to provide reliable and secure transmission. Due to increased network complexity risk is also increased and the system is vulnerable to threats and malicious attacks. An attacker can attack physical infrastructure resulting into power system failure, which may result into severe blackout in the region. An attacker can target the communication network by injecting false data resulting Denial of Service (DoS), which can lead to inaccurate meter measurements in smart metering applications.

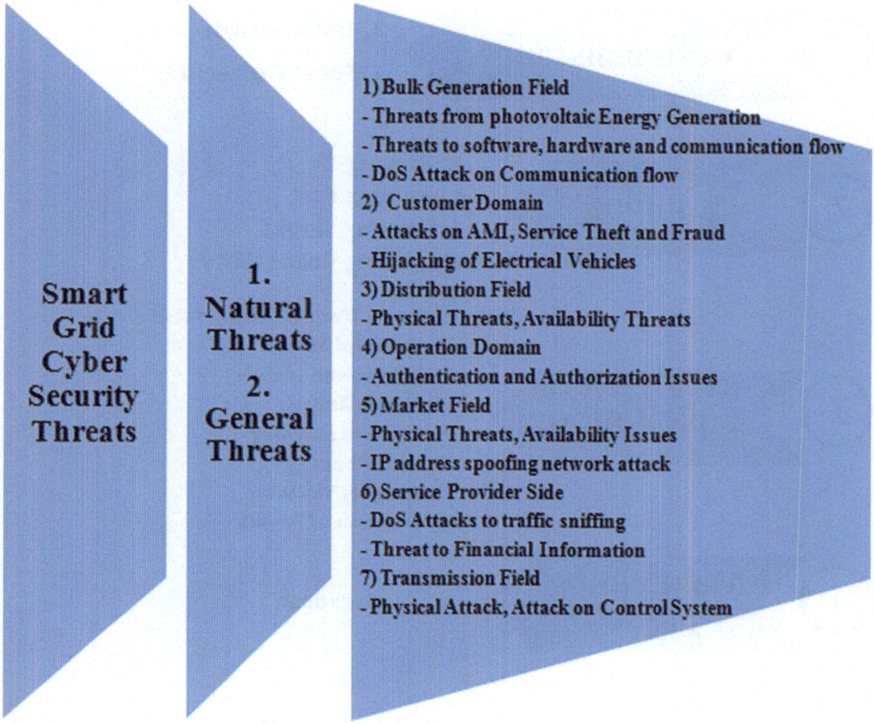

Fig. 10.3 Cyber security threats to smart grid CPS

Intruders can attack smart meter infrastructure by introducing vulnerabilities in smart meter measurements. An attacker can modify measured meter reading data by false data injection (FDI). The energy network makes possible to manage the distribution of energy in different forms such as power, LPG Gas, Water. A communication system controls power at source and consumption and meter measurements at the consumer end. Smart meter infrastructure (SMI) provides support for bidirectional meter data transmission. Smart meter reading data can be transmitted from meter to central control and from center through wireless media. The Proper secure protocol must be used for management of data and reliable data communication. Cyber networks comprise of communication technology used for monitoring and decision making necessary for effective data transmission (Habash et al. 2013).

The malign intent actors normally use sequential four phases to attack the smart grid system and to obtain the control over it. These phases are reconnaissance, scanning, exploitation and access maintenance as illustrated in Fig. 10.4. Information about the target system is assembled and collected in reconnaissance stage which is considered to be the inspection phase. System weaknesses are identified by the attacker in the scanning step. In the exploitation step, the invader tries to compromise and get control of the system. Once the victim system access is

Fig. 10.4 Cyber attacks in smart grid

obtained, attacker tries to maintain that access for permanent damage. During every stage, there is possibility of occurrence of various attacks on the system.

During reconnaissance stage, the attacks such as traffic analysis and social engineering are employed. Traffic analysis is a passive kind of attack which observes the traffic pattern and learns about the functioning of the system to find weaknesses of the devices, communication channels used, and hosts connected to the smart grid system. Under social engineering attack, malicious actor tries to win the trust of the victim user and employs communication tricks and encourages user to obtain credentials and personal information. In scanning attack, there is usually scanning of four things happen including IP addresses, open ports, associated services, and susceptibilities therein. Under exploitation stage, the weaknesses found and information gathered is used to attack the system and get control over it to apply further strategies. These attacks are of various types like viruses, worms, Trojan horses, etc. Other attacks of this kind are denial of service, man-in-the-middle attack, replay attack, channel jamming attack, HMI popping, violation of integrity and privacy used for making changes in the data stored in smart grid system. The attacker tries to obtain and retain permanent access to the target system by using backdoors, viruses, and Trojan horses under maintenance phase. In smart grid system, backdoor is embedded in the SCADA system for this kind of attack (El Mrabet et al. 2018).

10.5 Attack Detection Mechanisms

Cyber attacks have progressed from mere trouble intent to harm intents with the progress in wired and wireless communication, and control system innovative technologies. With the Internet growth, the number of sophisticated attacks has been increased. To detect and prevent this kind of attacks harming cyber physical systems is of paramount importance.

Among the shared responsibilities under cybersecurity paradigm, detection mechanism plays a vital role in deciding cyber security strategies to be applied (Cisco Midyear Cybersecurity 2016). There are five essential ways to detect the attacks as mentioned below:

Identification of strange emails from the unusual and unknown sources.
Noting of uncommon password related activities.
Recognizing doubtful pop-ups while browsing Internet.
Understanding and reporting of slower than normal Internet connections.
Keeping updated softwares on all the computing devices under use.

The authors in (El Mrabet et al. 2018) have proposed an attack detection strategy based on three phases such as pre-attack, under attack and post attack as mentioned in Table 10.2. None of the security system is full proof. Security is not a onetime thing, it is a continuous process. Preparedness for the cyber attacks come under pre-attack strategy. Network security keeps ready the intrusion detection system (IDS), security information and event management system (SIEM), network data loss prevention system and uses secure distributed network protocol version 3.0 (DLP3). Cryptography system has to keep the system well prepared through latest encryption, authentication and key management techniques. Under attack stage of detection strategy contains attack detection and attack mitigation mechanisms. Attack detection strategy consists of IDS, SIEM, DLP, and various data stream mining based IDS such as accuracy updated ensemble, active classifier, leveraging bagging, limited attribute classifier, bagging using ADWIN and adaptive size tree, and single classifier drift. For attack mitigation, pushback and reconfiguration method is used for DoS attack and anti-jamming schemes are used for jamming attacks. Post attack mechanisms generally employ forensic analysis that analyzes IDS signature, tresses in anti-virus database and applied security policies.

10.6 Cyber Security in Smart Grid

Physical infrastructure is vulnerable to attacks by terrorists; these attacks may result in power failure in some critical region. Power outage in the central control system results in cascade power failure and may result in a blackout to a great. Some spying agents are trying to access data for modification. Due to lack of skills human errors occur. Technical errors may occur and result in the failure of one or more

Table 10.2 Smart grid cyber attack detection techniques (El Mrabet et al. 2018)

Situation	Category	Attack detection techniques
Pre-attack (Preparedness for any attack)	Network security	Intrusion Detection System (IDS)
		Security Information and Event Management System (SIEM)
		Network Data Loss Prevention (DLP)
		Secure Distributed Network Protocol Version 3.0 (DNP3)
	Cryptography	Encryption
		Authentication
		Key management
	Device security	Host IDS
		Compliance checks
		Diversity technique
Under attack	Attack detection	SIEM, DLP, IDS
		Data stream mining based IDS • Accuracy updated ensemble • Active classifier • Leveraging bagging • Limited attribute classifier • Bagging using ADWIN • Bagging using adaptive size Hoeffding tree • Single classifier drift
	Attack mitigation	For DoS attack • Pushback • Reconfiguration method
		For jamming attack • Anti-jamming schemes using fuzzy logic
Post attack	IDS signature	
	Anti-virus database	
	Security policies	
	Forensic analysis	

devices in power generation plants. Due to increased complexity of supervisory control and data acquisition (SCADA) system, it adds the risk of threats to physical and cyber infrastructure.

Smart grid vulnerabilities introduce attacks on physical components of power systems. Cyber infrastructure is targeted by viruses and threats cause failure of software monitoring and controlling smart grid operations. Control centers in all regions are interconnected and form grid network for communication and control of smart grid services. Typical anomalies found in smart grid are False Data Injection (FDI) and Denial of Service (DOS). The intruder can modify data by introducing false data which can seriously affect the integrity of data. Denial of service prevents access to services to the authorized user which effects on data availability (McDaniel and McLaughlin 2009).

10.6.1 Smart Grid Cyber Security Needs

For security achievement, the main parameters are Confidentiality, Integrity, Availability and Accountability in cyber security as shown in Fig. 10.5. Data privacy and security are very much relevant parameters in Smart Grid communication networks. Lot of research work is going on different aspects of smart grid including efficiency, reliability, scalability, privacy, latency.

Confidentiality is very much important for smart grid is related to meter data management. Appropriate meter control, usage of meter, and unaltered correct billing information is very much necessary for the successful smart grid operations. The information communication flow through channels can be changed or the channels can be jammed thereby making them unavailable to the system data transactions. Integrity of the actual data related to energy transfer in full duplex mode should be maintained by the power injection meters and state estimators. Consumer monthly electricity bills should be accountable with actual usage and they should not have been changed in between. Security need and targets to be protected are explained in Table 10.3.

10.6.2 Related Work

Probabilistic risk analysis framework for smart grid is presented in the research work of (Smith and Pat′e-Cornell 2018) which helps in deciding the requirement of smartness level, connectivity level and associated risks. The authors have proposed a multi-armed bandits (MAB) approach based on Bayes adaptive network security model. Cyber defense teams among the network node entities to work in proactive mode are used to for information collection and security purpose with maximum connectivity. In the work of (Babun et al. 2016), a configurable framework for detection of forged smart grid devices is presented. It uses system call tracing, library interposition, and statistical procedure for monitoring and detection of

Fig. 10.5 Cyber security requirements

Table 10.3 Security needs and targets to be protected

S. no.	Security needs of smart grid	Target to be protected against cyber attacks
1	Confidentiality	Control of a meter, metering usage, billing information
2	Availability	Obstructing the information flow through network and making smart grid network unavailable
3	Integrity	Power injection meters and power flow to the state estimator
4	Accountability	Monthly electricity bills through smart meters

malicious devices in the system. Especially two categories are taken into consideration like resource-rich and resource-limited smart grid devices. Realistic testbed for smart grid is used with GOOSE messages with IEC-61850 communication protocol. This framework is detecting malicious actors in smart grid with good pace.

The researchers have presented an attack on the electric metering system in (Wu et al. 2019). The attack is able to make considerable changes in the calculated electricity utilization. Here, the power circuit is switched on/off depending on meter's sampling rate by the malicious actor. This false load attack is so much sophisticated so that it is immune to the time stamped secure channel made up by using cryptographical system. The eavesdropper can compromise controllers present in the grid system. Meter and communication channel are untouched by the attacker because of which it remains undetected. In contrary to this, false load injection attack can be detected by standard security solutions which targets either meter or communication channel.

Game theoretic model based attacker defender system is developed and analyzed in the work by (Sanjab and Saad 2016). Cognitive hierarchy theory inspired bounded rationality framework suitable for limited level thinking attacker is introduced in this paper. The designed framework is applied to the smart grid and energy market proposition. Defender shows better protection of the smart grid system in limited level thinking attacker. It is observed that the gain from Nash Equilibrium defense system decreases if the number of thinking level of the attacker is increased. Hardware software codesign framework for smart grid hardware-in-the-loop (HIL) is presented in paper (Albarakati et al. 2018). The HIL testbed is a combination of a real time power grid simulator and an open stack based communication network. This co-simulation framework is capable enough to respond to the attacks on its power and communication counterparts. Various attacks were imposed on this co-simulation framework and corresponding behavior of the smart grid system was observed. This co-simulation framework is proficient of simulation of different smart grid protocols including GOOSE, Sampled Values (SV), PTP, IEEE C37.118, IEC 61850-90-5, DNP3, and Modbus.

In research work of (Lei et al. 2018), cyber physical security and cyber physical reliability fears are elaborated with their capacities, approaches, and advancements.

Smart grid is progressing well with the growth of information and communication technology (ICT). Due to the amalgamation of electric grid and ICT, improvements are obvious in system monitoring, control, protection, and data processing competency. But cyber attacks on these systems considerably degrade the overall performance of the smart grid system. It affects on the reliability of the communication among different entities of smart grid infrastructure. The work in (Xiang and Wang 2019) addresses the mathematical modeling and solution algorithm development for multi-attack scenario defender-attacker-defender for smart grid. Attacker's nasty resource suspicions are detained. This model is crumbled into two parts such as upper level problem (ULP) and lower level problem (LLP). Stochastic programming mechanism is applied in ULP. This defender-attacker–defender model can reduce the probable load loss in the multi attack scenario. Real time security approach for the smart meters in smart grid environment against external attacks in a distribution substation is projected in (Buyuk 2018). Message passing interface (MPI) system is used with the Floyd–Warshall algorithm for distribution of key pairs generated by homomorphic encryption. Raspberry Pi2 model B is used for implementation purpose. History of power grid system attack shows that the attacks were long lasting. This type of parallel program approach significantly solves the attacks in millisecond period of time.

A micro grid structure with appropriate generation and distribution specifications and significant communication infrastructure is designed for analysis of security challenges in the smart grid is presented in (Jahan and Habiba 2015). For cyber physical system like smart grid, it is very crucial to provide both information security and system theory based security. Two events including multimode event and dispatch event are considered while developing the structure. Power factory 14.1 software package is used for simulation performance.

10.6.3 Smart Grid AMI Security Techniques

Smart grid security technique is a mechanism to provide security with privacy for physical infrastructure and data communication. To ensure privacy, secret keys can be established prior to the communication process. Routing protocol attacks may disturb complete connectivity among smart grid devices. The security protocols must be designed considering the requirements of different modules of AMI communication network such as Home Area Network (HAN), Neighborhood Area Network (NAN) and Wide Area Network (WAN) serving numerous applications of smart grid. There are different security techniques available for smart grid considering parameters such as false data injection (FDI), survivable to intrusions, end-to-end access protection and delay sensitivity, power quality and voltage issues.

Methods available for smart grid security are as follows

(1) State estimation Techniques.
(2) Framework design.
(3) A Secure key management scheme or cryptographic algorithm.
(4) Security protocol design with error detection.

(1) **State Estimation Techniques**
Technique in which meter reading data is processed to estimate unknown state variables is called as state estimation techniques. The state estimation output is used to control the power flow and detection of faults. It also prevents false data injection attack and monitors quality of power by comparing current state with previous. State estimation model considers state of parameters such as voltage, current, phase etc., thus by solving the linear equations for all of the measurements it estimates current state. Attacker may attack on the basis of knowledge of some parameters which are known, but state estimation model considers all parameters and by comparing current and past state these attacks can be easily detected.
(2) **Framework Design**
The smart grid is next generation power system which provides bidirectional communication and serves various applications. Smart grid consists of two infrastructure i.e. physical and cyber infrastructure; therefore it is important to consider security and reliability of smart grid infrastructure. From security perspective framework model design approach can be used which considers various parameters. Framework is combination of different structure, methodology and designs, all these when framed together considering different parameters. Framework approach is useful for detection and detection and prevention from various threats to smart grid infrastructure.
(3) **A Secure key management scheme or cryptographic algorithm**
A secure key management scheme is available for secure data communications in a smart grid system. Key management scheme uses either public key or private key algorithms to encrypt and decrypt data. In public key algorithm both sender and receiver uses same key to encrypt and decrypt data, whereas in private key algorithm sender and receiver uses different keys to encrypt and decrypt data. Time required to decrypt or encrypt data depends on key length. In smart grid to provide secure data communication data various public key or private key algorithm are used.
(4) **Security protocol design with error detection**
Smart grid provides bidirectional communication, hence many smart applications uses smart grid network for data communication. It is important to provide secure communication and data transfer for these applications. Therefore it is important to design a communication protocol for secure and reliable communication. These communication protocols will have capability to detect and prevent attacks to smart grid networks.

10.7 Summary

It was possible for conventional power grid system, to keep them safe by protecting behind fences and locked doors. But with today's smart grid, it has huge physical cyber assets which cannot be secured only with fences and locks. It needs intelligent and effective cybersecurity provisions with assured privacy and trust. AMI systems which not only reduce cost but also provide other benefits at customer and operational level such as, personalize tariff plans, improved service quality, and easy bill options should possess robust security provisions for theft intrusion detection. Electricity grids are national assets and they must be utilized with due care and safety measures. SCADA is the backbone of communication system in the smart grid. It is very important and challenging task to keep it secure. Attack surface is increasing with the expansion and adoption of smart grid. The malicious actors who seek to harm smart grid continue to grow in intelligence, number, and level of commitment. In the same or more proportion, the security defense mechanisms should be designed and developed. Lot of research is needed to make Smart Grid system more strong, secure, reliable and cost effective.

References

Albarakati A, Moussa B, Debbabi M, Youssef A, Agba BL, Kassouf M (2018) OpenStack-based evaluation framework for smartgrid cyber security. In: IEEE international conference on communications, control, and computing technologies for smart grids (SmartGridComm), 2018, pp 1–6

Babun L, Aksu H, Uluagac AS (2016) A framework for counterfeit smart grid device detection. In: IEEE conference on communications and network security (CNS), IEEE CNS 2016

Bartoli A, Hernandez-Serrano J, Soriano M, Dohler M, Kountouris A, Barthel D (2010) Secure lossless aggregation for smart grid M2M networks. In: 2010 first IEEE international conference on smart grid communications, 2010, pp 333–338

Baumeister T (2011). Literature review on smart grid cyber security. In: Semantic scholar

Berthier R, Sanders WH, Khurana H (2010) Intrusion detection for advanced metering infrastructures: requirements and architectural directions. In: 2010 first IEEE international conference on smart grid communications, 2010, pp 350–355

Buyuk OO (2018) A novel actual time cyber security approach to smart grids. In: 6th international Istanbul smart grids and cities congress and fair (ICSG), 2018

Cisco Midyear Cybersecurity Report on "5 Ways of Detecting A Cyber Attack" (2016). https://www.cisco.com/c/dam/m/en_ca/business-transformation/pdf/5-ways-to-detect-a-cyber-attack.pdf

Dan G, Sandberg H (2010) Stealth attacks and protection schemes for state estimators in power systems. In: 2010 first IEEE international conference on smart grid communications, 2010, pp 214–219

Efthymiou C, Kalogridis G (2010) Smart grid privacy via anonymization of smart metering data. In: 2010 first IEEE international conference on smart grid communications, 2010, pp 238–243

El Mrabet Z, Kaabouch N, El Ghazi H, El Ghazi H (2018) Cyber-security in smart grid: survey and challenges. Science Direct Elsevier J Comput Electr Eng 67:469–482

Habash RWY, Groza V, Krewski D, Paoli G (2013) A risk assessment framework for the smart grid. In: IEEE Electrical Power and Energy Conference (EPEC), 2013

Jahan S, Habiba R (2015) An analysis of smart grid communication infrastructure and cyber security in smart grid. In: Proceedings of 3rd international conference on advances in electrical engineering, 2015

Jiaxi Y, Anjia M, Zhizhong G (2006) Cyber security vulnerability assessment of powerindustry. IEEE PES power systems conference and exposition, pp 2200–2205

Kalogridis G, Efthymiou C, Denic SZ, Lewis TA, Cepeda R (2010) Privacy for smartmeters: towards undetectable appliance load signatures. In: 2010 first IEEE international conference on smart grid communications, 2010, pp 232–237

Khurana H, Bobba R, Yardley T, Agarwal P, Heine E (2010) Design principles for powergrid cyber-infrastructure authentication protocols. In: Hawaii international conference on system sciences, 2010

Kosut O, Jia L, Thomas RJ, Tong L (2010) Malicious data attacks on smart grid stateestimation: attack strategies and countermeasures. In: 2010 first IEEE international conference on smart grid communications, 2010, pp 220–225

Lei H, Chen B, Butler-Purry KL, Singh C (2018) Security and reliability perspectives in cyber-physical smart grids. In: IEEE innovative smart grid technologies-Asia (ISGT Asia), 2018, pp 42–47

Li H, Lai L, Qiu RC (2010) Communication capacity requirement for reliable and securestate estimation in smart grid. In: 2010 first IEEE international conference on smart grid communications, 2010, pp 191–196

Li F, Luo B, Liu P (2010) Secure information aggregation for smart grids using homomorphic encryption. In: 2010 first IEEE international conference on smart grid communications, 2010, pp 327–332

McDaniel P, McLaughlin S (2009) Security and privacy challenges in the smart grid. IEEE Secur Priv Mag 7(3):75–77

NIST (2010) Guidelines for smart grid cyber security: vol 2, privacy and the smart grid. The Smart Grid Interoperability Panel—Cyber Security Working Group. In: Homeland Security Digital Library, 2010

Olmstead S, Siraj A (2011) Smart grid insecurity: a new generation of threats. In: Proceedings of the international conference on security and management (SAM'11), Las Vegas, NV, 18–21 July 2011

Rahman MA, Al-Shaer E, Rahman MA (2013) A formal model for verifying stealthy attacks on state estimation in power grids. In: IEEE symposium on smart grid, cyber security, and privacy, SmartGridComm 2013

Sanjab A, Saad W (2016) On bounded rationality in cyber-physical systems security: game-theoretic analysis with application to smart grid protection. In: Joint workshop on cyber-physical security and resilience in smart grids (CPSR-SG), 2016, pp 1–6

Smith MD, Pat´e-Cornell ME (2018) Cyber risk analysis for a smart grid: how smart is smart enough? a multiarmed bandit approach to cyber security investment. IEEE Trans Eng Manag 65(3):434–447

So HKH, Kwok SH, Lam EY, Lui KS (2010) Zeroconfiguration Identity-based signcryption scheme for smart grid. In: 2010 first IEEE international conference on smart grid communications, 2010, pp 321–326

Valdes A, Cheung S (2009) Intrusion monitoring in process control systems. In: Proceedings of the 42nd annual Hawaii international conference on system sciences HICSS, 2009, pp 1–7

Varodayan DP, Gao GX (2010) Redundant metering for integrity with information theoretic confidentiality. In: 2010 first IEEE international conference on smart grid communications, 2010, pp 345–349

Vukovic O, Dan G (2013) Detection and localization of targeted attacks on fully distributed power system state estimation. In: IEEE symposium on smart grid cyber security and privacy, martGridComm 2013

Watts D (2003) Security and vulnerability in electric power systems. In: 35th North American power symposium 2003, pp 559–566

Wu Y, Chen B, Weng J, Wei Z, Li X, Qiu B, Liu N (2019) False load attack to smart meters by synchronously switching power circuits. IEEE Trans Smart Grid 10(3):2641–2649

Xiang Yingmeng, Wang Lingfeng (2019) An improved defender-attacker-defender model for transmission line defense considering offensive resource uncertainties. IEEE Trans Smart Grid 10(3):2534–2546

Xie L, Mo Y, Sinopoli B (2010) False data injection attacks in electricity markets. In: First IEEE international conference on smart grid communications, 2010, pp 226–231

Zhang J, Gunter CA (2010) Application-aware secure multicast for power grid communications. In: 2010 first IEEE international conference on smart grid communications, 2010, pp 339–344

Chapter 11
Bluetooth Communication

While living in this smart and intelligent world, many times, consumer connects to some or the other device through Bluetooth. This includes HVAC system, televisions, smart phones, fitness trackers, connected cars, Internet of things, connection of wireless mouse to laptop, wireless headphone connection to mobile device, etc. Wirelessly connected gadgets have become integral part of human life. A mobile malware Lasco. A started targeting to mobile devices in 2005. It uses Symbian operating system and uses Bluetooth technology to replicate itself and also gets spread to other Bluetooth enabled nearby devices. The infection can render the mobile device unstable (Minar and Tarique 2012). The primary cause of spreading of this disease was Bluetooth enabled systems. The worms always cause some harm to the network. Being a low cost and ad hoc technology blue tooth is used by almost every user for communication. But while communicating using Bluetooth no one has thought of serious attacks like stealing your personal information like your PIN of money transfer or any other information. Recent news on January 24th, 2014 was "A criminal organization hit gas station ATMs located in South America. The gang used Bluetooth-enabled skimmers to steal two million dollars from customers". Here the hackers made use of Bluetooth technology for hacking the PIN of an account holder (Doon 2014).

The wireless technologies are responsible for considerable changes in networking and telecommunication services by integrating the networks. These techniques make use of the shared medium, limited resources, etc. Hence, the wireless ad hoc networks are vulnerable to a variety of potential attacks. Governments, military, financial institutions, hospitals and other businesses assembly process and amass a large amount of confidential information on computers, transmit this information across networks. Not only these large organizations but a person also sends the information across the network by making use of the internet or any wireless technology. But nowadays this transmission of information is not secure because some people are trying to use your personal or confidential information in an illegal way which is a crime.

R. Prasad and V. Rohokale, *Cyber Security: The Lifeline of Information and Communication Technology*, Springer Series in Wireless Technology,
https://doi.org/10.1007/978-3-030-31703-4_11

Bluetooth is also a means of wireless communication, and hence, any illegal activity done by making use of Bluetooth communication comes under the cyber-crime. The name 'Bluetooth' was given in the memory of Viking king in Denmark Harald Bluetooth who brought together and controlled Denmark and Norway. Bluetooth is a powerful wireless technology that is used for formation of small networks in short time span and it does not need any infrastructure. Because of its adhoc nature, low power requirement, and no need to use cables, and with low cost, Bluetooth has become widely adopted leading technology.

11.1 Introduction

Ericsson Mobile Communications initiated research on universal short-range and low-power wireless connectivity primarily for exchanging data between mobile phones, headsets, vehicular devices, and computers which ended in a group known as Bluetooth Special Interest Group (SIG).Bluetooth technology was officially granted in 1999 (Minar and Tarique 2012). Then it was adopted in various electronic devices. Figure 11.1 shows the different steps of development of IEEE802.11.

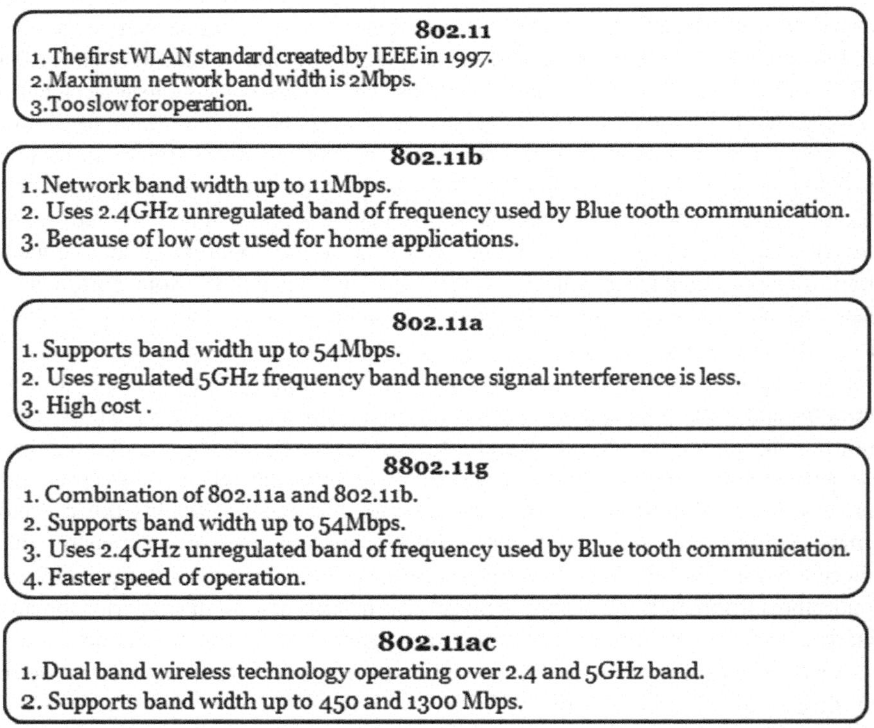

Fig. 11.1 Evolution of IEEE802.11

Existing Bluetooth technology operates over the 2.4 GHz Industrial Scientific Medical (ISM) unlicensed frequency band which is mainly for low power communications. Within 10–100 m, Bluetooth can provide 700 Kbps, 2.1 Mbps, or up to 24 Mbps data rates depending on its version. One thing that has become apparent is that Bluetooth security and user ease rivets some serious tradeoffs that is why the attacks are more frequent in this type of communication.

The security system in Bluetooth builds upon three procedures: Pairing, Authentication, and Encryption. Presently modified SAFER+ algorithm is used for higher data throughput and frequency and is used for authentication and encryption; it is having a limitation of Encryption/Decryption, dissimilarity, and no proven security. The Triple DES and Tiger algorithm and Identity-based cryptography are the mechanisms used for secure encryption. This research work proposes a Bluetooth security mechanism making use of Galois field and encryption (Rajput et al. 2012). The chapter is organized as follows. Section 11.2 discusses the correlation in between Bluetooth and cybercrime. Attacks on Bluetooth communication are elaborated in Sect. 11.3. Section 11.4 throws light on state of the art in Bluetooth Security techniques. Chapter is summarized in Sect. 11.5.

11.2 Bluetooth and Cyber Crime

There was news on the internet which said “A criminal organization hit gas station ATMs located in South America. The gang used Bluetooth-enabled skimmers to steal two million dollars from customers”. The Bluetooth-enabled skimmers used are a small device, able to capture credit card data. This information was used to clone customer’s card and use them to draw cash from ATMs in other states. This news was one of the inspirations to work on Bluetooth security vulnerabilities. When we hear the word Bluetooth we do not think it as a far-reaching crime but after reading the news, we will come to know the seriousness.

The name ‘Bluetooth’ was given in the memory of Viking king in Denmark Harald Bluetooth. It was formed by the Ericsson in 1994, it was designed to replace RS-232 data cables with wireless channels. In the development journey of Bluetooth, there are different versions which are described as below. All versions of the Bluetooth standards are designed for downhill compatibility. The latest standard covers all previous versions. The characteristics of different versions are described in the sections as follows.

11.2.1 Bluetooth v1.0 and v1.0B

The evolution of different Bluetooth versions plays the critical role in the understanding of this Bluetooth communication. Bluetooth hardware device address (BD_ADDR) is the fixed part of the versions 1.0 and 1.0B for the connection

Table 11.1 Description of different Blue tooth versions (Doon 2014)

Bluetooth versions	Specifications
v1.1	1. Accepted as IEEE Standard 802.15.1in the year 2002 2. Added possibility of non-encrypted channels
v1.2	1. Speedy detection and connection 2. This version makes use of adaptive frequency hopping spread spectrum (AFHSS) which has good resistance to radio frequency interferences 3. Quite high transmission rates up to 721 Kbps 4. Standardized as IEEE 802.15.1 in the year 2005 5. Flow control and retransmission modes for L2CAP
v2.0 + EDR (Enhanced Data Rate)	1. Released in 2004 2. Faster data transfer and reduced power consumption
v2.1 + EDR (Enhanced Data Rate)	1. Adopted on 26 July 2007 2. Use of secure simple pairing (SSP) 3. Included "Extended Inquiry Response" 4. Made use of sniff subtracting, which reduces the power consumption in low-power mode
v3.0 + HS (High Speed)	1. Approved on 21 April 2009 2. Data transfer speeds of up to 24 Mbit/s 3. The main new feature is AMP (Alternative MAC/PHY), the addition of 802.11 as a high speed transport
v4.0	1. This version was completed as Bluetooth Smart on 30 June 2010 2. Includes Classic Bluetooth, Bluetooth high speed and Bluetooth low energy protocols
v4.1	1. Announcement of adoption of the version on 4 December 2013 2. Improved consumer usability, bulk data exchange rates, and allowed devices to support multiple roles simultaneously

process aimed at data transmission. The different versions and their characteristics are described in the following Table 11.1.

11.2.2 Key Steps for Bluetooth Security

Here different key steps for Bluetooth security such as authentication, authorization (pairing), and encryption regarding confidentiality have been described.

1. **Authentication**
   The procedure of authentication is used for identity verification. The Bluetooth device authentication procedure is a challenge-response method. Each device in this system is referred to as either the claimant or the verifier. The candidate is trying to prove its identity, and the verifier is confirming the identity of the provider. This protocol authenticates the electronic device by verification of the Bluetooth link key which is basically secret key (Vishwakarma and Patel 2011).

2. **Authorization/Pairing**
   Two electronic devices wishing to connect with each other for data transaction need to go through pairing process. A common key for authentication and encryption is generated in between two Bluetooth devices which have set up the connection for data transmission and reception.
3. **Encryption/confidentiality**
   Encryption is a vital part in the steps of Bluetooth security. In the procedure of encryption, the information is encoded in such a way that an attacker will not be able to decode its contents. A random number must be sent between the two devices who wish to communicate with each other. The receiving end device should also know the PIN of the sending end devices. With the help of these two steps, a connection key is generated on both devices, and we can say that the communication is now confidential from the hacker (Patheja et al. 2011).

11.3 Attacks on Bluetooth Communication

There are various types of attacks/vulnerabilities present in Bluetooth system. All these attacks are discussed in brief in Table 11.2 shown.

The attacks are related to key generation, encryption and with authentication. For example, the attacks which are associated with the entire production or Personal identification number (PIN) are Media access control (MAC) spoofing attack, PIN cracking attack. Mac spoofing attack will come in the process of encryption also. It is not necessary that the attacks will relate only a single parameter from encryption, authentication, and pairing (Hassan et al. 2018). The Same attack may cover various parameters.

11.3.1 Identifying Types of Attacks

There are various types of Bluetooth vulnerabilities which are called as Bluetooth attacks. A better understanding of weaknesses and attacks can be achieved by grouping them based on shared properties and similarities. These similarities are based on the steps of Bluetooth security. This can be explained with the help of Fig. 11.2 which is shown. Authentication related attacks can be listed as blue snarfing, blue jacking, blue printing, man-in-the-middle, blue bug, etc. Attacks associated with pairing mechanism include blue bug, PIN cracking, man-in-the-middle, blue over, offline PIN recovery, reflection, backdoor, and denial of service (Sharma 2016). Encryption molest types contain MAC spoofing, cabir worm, skulls worm, Lasco worm, etc.

Table 11.2 Different types of attacks on Bluetooth communication (Minar and Tarique 2012)

S. no.	Bluetooth attacks	Description
1	MAC spoofing attack	1. Occurs during link key generation 2. With the use of special hardware, attacker use spoofing to terminate valid connections and/or manipulate data while in transit
2	PIN cracking attack	1. The most frequently reported attacks 2. The attacker tries to find the PIN to get the correct initialization key
3	Man-in-the-middle/ impersonation attack	1. Attacker receive the message between two communicating devices and pass on the message 2. Makes a false assumption for the communicating devices that they are paired where actually they are paired with the attacker
4	Blue jacking attack	1. Voluntary messages are sent to the Bluetooth enabled devices which are available inside the range of 10 m 2. Do not alter any data
5	Blue sniffing attack	1. Attacker gain unauthorized access to Bluetooth enabled mobile phone and can copy anything which is stored in the phone memory 2. Software tools are required to manage this attack
6	Blue bugging attack	1. Attacker connects to the target phone in unaware of its owner 2. Attacker can send text messages to premium numbers initiate phone calls to premium numbers, write to phonebook entries and many more
7	Blue printing attack	1. To get information about the manufacturer, device model and firmware of the target device, it is used 2. This information is used for finding out vulnerabilities and then attacking the target device
8	Blue over attack	1. Makes use of blue bugging attack 2. Can be done by using only Bluetooth mobile phone with Blue over or Blue over II installed
9	Off-line PIN recovery attack	Guessing of different PIN values are guessed and then sensitive information is obtained
10	Brute-force attack	1 A brute-force method is used
11	Reflection attack	1. The attacker only reflects the received information from one target device to another during the authentication
12	Backdoor attack	1. Trust relationship is established through pairing process 2. Without the owner's knowledge, sensitive information is gathered to access the Internet, WAP, and GPRS gateways
13	DoS attacks	1. Attacker attacks against the physical (PHY) layer to jam the complete Piconet
		3. Attacker sends random data in every timeslot
14	Cabir worm	1. It is malicious software which finds available Bluetooth devices and sends itself to them

(continued)

Table 11.2 (continued)

S. no.	Bluetooth attacks	Description
15	Skulls worm	1. Symbian Installation System (SIS) Trojan file that pretends to be Macromedia Flash player is used for sophisticated attacks 2. User has to open the file and install it, after activation it starts to search for new target
16	Lasco worm	1. It is a SIS file infecting virus which enters into the SIS file and infects it

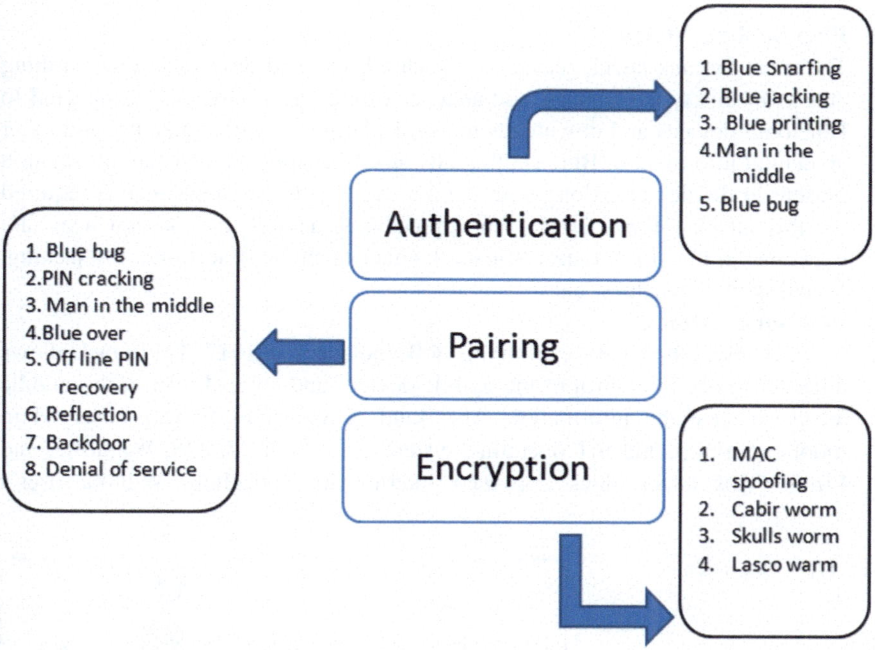

Fig. 11.2 Classification of Bluetooth attacks

11.3.2 Classification of Bluetooth Attacks

Out of these types, the most frequent attacks are blue bug attack, blue jacking attack, and blue Sniffing attack. These attacks are described as below:

1. **Blue Jacking Attack**
   Similar to email spam and phishing, in Blue jacking attack, the unauthorized user can send unsolicited messages, or business cards, to Bluetooth-enabled devices as shown in Fig. 11.3.

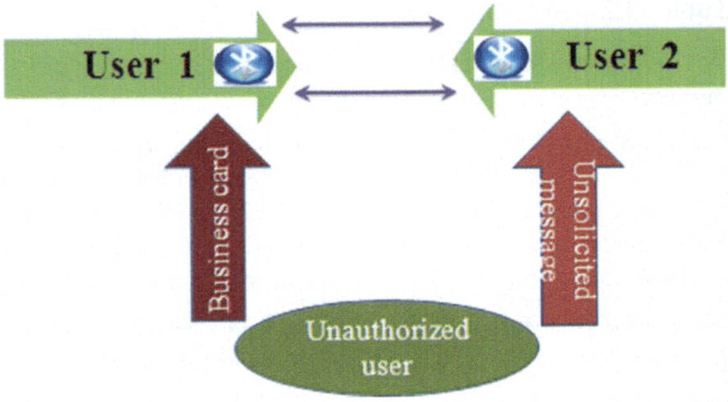

Fig. 11.3 Blue jacking attack

2. **Blue Sniffing Attack**
The most famous attack related to Bluetooth enabled devices is blue Sniffing attack. In this type of attack, the attacker usually gets wirelessly connected to Bluetooth devices and downloads the stuff like phonebook, calendar, and other credential information. Blue sniffing advanced version can alter the information present in the device. Along with the hardware, software assistance is required for this attack (Henda 2014). The procedures, as well as consequences, are explained in Fig. 11.4. Latest Bluetooth attacks include BlueBorne, Btlejacking, Bleedingbit, CarsBlues, etc.
3. **BlueBorne Attack**
Security researchers have explored BlueBorne attack in 2017. This attack allows a hacker to obtain control of the mobile devices and applied man-in-the-middle attack to steal the information. This kind of vulnerability was observed in mobile, desktop, and IoT operating systems like Android, iOS, Windows, and Linux. This attack does not need Internet for spreading. Without user's

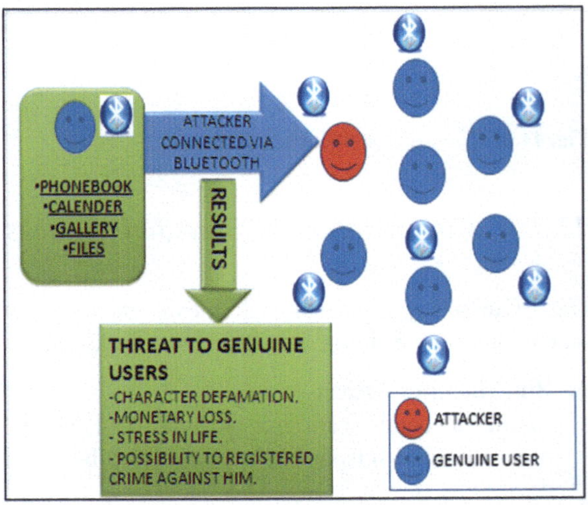

Fig. 11.4 Blue sniffing attack

knowledge, the hacker can get connected quietly to the target device and takes control of it to launch next attacks.

4. **Btlejacking Attack**
   This attack was detected in 2018 by Damien Cauquil in Las Vegas conference. Btlejacking attack can jam and takeover any Bluetooth low energy device (BLE). For the attack to be successful, the attacker and hacker should be in five meters range from each other. IoT devices use BLE protocol. There is potential threat to IoT resource constrained devices due to this kind of attacks.
5. **Bleedingbit Attack**
   Two Bluetooth chip vulnerabilities are identified by the researchers such as CVE-2018-16986 and CVE-2018-7080. With a remote code execution error, attackers can remotely send malicious BLE broadcast messages under the name of advertising packets and can lead to memory exhaustion. These vulnerabilities can allow an attacker to access and install totally new and strange version of the firmware.
6. **CarsBlues Attack**
   Infotainment system of several vehicles was found infected by CarsBlues attack. Personal information is erased from the hardware and software of the vehicle. Millions of vehicles are suffered due to this kind of attack. This attack can lead to vehicle hijacking or it may put human life into danger by leading to road accidents (CYWARE 2019).

11.4 State of the Art in Bluetooth Security Techniques

Security level of data transmission during Bluetooth communication IS expected to be very high. The present security mechanisms make use of DES and RSA Algorithm for encryption of data. DES algorithm is used to encrypt the Data and RSA algorithm is used to encrypt the Key. RSA Algorithm uses careful factoring; its computing velocity is slower than IDEA, and it is suitable for encrypting a small amount of data. IDEA Algorithm is suitable for encryption of a vast number of data. The RSA algorithm can distribute encryption key openly, it is also very easy to update the encryption keys, and for the different communication, objects compared to IDEA. The combination of existing available security solutions can be a good choice as put forth in (Rajput et al. 2012).

The E0 stream cipher algorithm is used for encryption in Bluetooth, but this algorithm does not have enough data security. Here the authors have proposed an Integrated Encryption Scheme based on IDEA, RSA and MD5 algorithms. The IDEA is a symmetric algorithm which operates on 64-bit plaintext blocks by making use of 128-bit keys, giving rise to a system which is practically more immune to brute force attacks.

In general Triple DES uses three independent keys called as the principal bundle. In (Vishwakarma and Patel 2011), the authors have proposed a hybrid encryption technique using triple DES for encryption of the key with the use of Tiger algorithm. In Tiger algorithm, there is a double protection of data so that we can use Bluetooth technology more widely. The Existing SAFER+ algorithm is modified to provide higher data output and frequency with the help of some modifications like an introduction of Rotation block between every round in (Hassan et al. 2018). The author has used some of the advantages of SAFER+ algorithm like its Speed and Simplicity, Transparency, Flexibility of Use, etc.

In Bluetooth communication encryption is one of the necessary steps. Present technologies use E0 encryption algorithm. But the algorithm has the disadvantage of slow operational speed which may help the attacker to break the encryption along with that it consumes more battery life of Bluetooth devices. These difficulties may be overcome by using an RC4 algorithm. In the research work of (Mora-Afonso et al. 2013), the authors have proposed a scheme for the use of RC4 algorithm by utilizing some of its advantages like the difficulties of knowing values in the Table, The difficulties of identifying the location in the Table used to select each value in a sequence. Hence, the encryption speed is faster compared to other techniques.

In research work by Latchmanan and Parveen (2018), the author has proposed Identity-Based Cryptography as a method for implementing secure, useful and efficient schemes in smart mobile devices. However, data-leaking is one of the serious and frequently arising phenomena in the use of Bluetooth technology for data transfer. Hence, if we use RSA algorithm and the DES key for data transmission, there is no need to transfer DES key secretly before communication, and DES algorithm is used for data transmission because of its higher efficiency in block encryption, RSA algorithm is used for the encryption of the key of the DES because of its advantages in the major cipher. In the research work of (Mora-Afonso et al. 2013), the authors have proposed a scheme of blue tooth security using DES and RSA hybrid encryption algorithm and have shown that it is relatively more secure, thus ensuring data transmission between the Bluetooth device safety and real-time.

Bluetooth wireless technology is an inexpensive, short-range radio technology that does not require cabling between devices allowing an exchange of valuable data or files, image, messages, personal videos, etc. Blue bug attack is one of the dangerous type of Bluetooth attack where a mobile phone can be damaged or lost valuable information and attacker can make calls, send messages, read phone books, examine calendars, etc. When a user put balance in mobile, it can reach zero by blue bug attack. RFCOMM is a direct connection between two devices and Bluetooth RFCOMM has no authentication. With this RFCOMM, attackers very easily enter the new device and establish the direct link without authentication. In (Reddy and Raju 2013) the authors have proposed authentication approach for protection of blue bug attack. For this, authors have suggested that RFCOMM should use a database for Bluetooth device address or Mac address. After scanning, when the match is found then only RFCOMM should give permission to communicate the particular appliance, otherwise it should not get permission. All these steps are illustrated in Fig. 11.5.

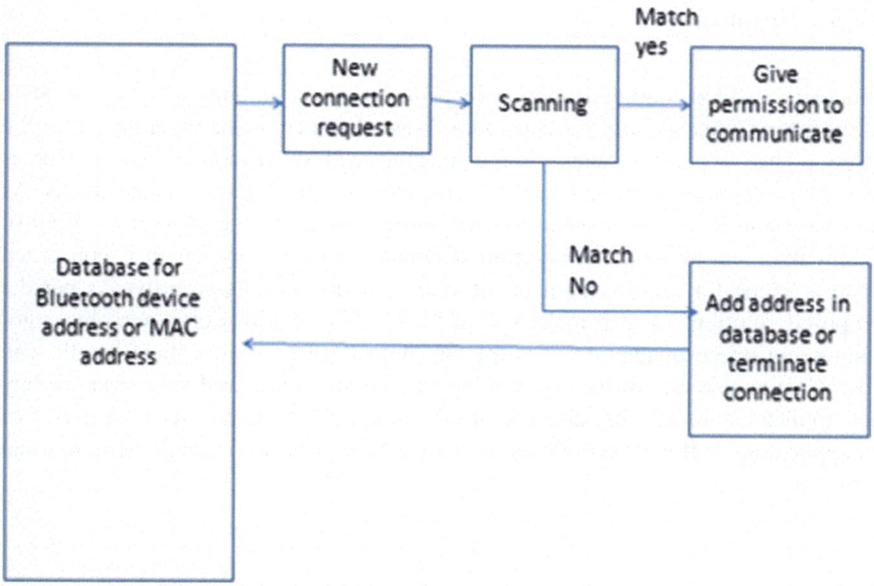

Fig. 11.5 Procedure of RFcomm authentication function

As per the basic steps required for Blue tooth communication and the previous work done in this field we have summarized all the techniques which are used till now for the safe use of Bluetooth exchanges in Table 11.3.

Table 11.3 Mechanisms used for Bluetooth security

Ref. paper	Authentication	Pairing	Encryption
(Rajput et al. 2012; Vishwakarma and Patel 2011; Latchmanan and Parveen 2018)	IDEA + RSA + MD5	RSA, BTDA	RC4
(Vishwakarma and Patel 2011; Reddy and Raju 2013; Henda 2014; CYWARE 2019)	RSA, BTDA	SAFER + Cipher algorithm for MITM	DSE + RSA
(Patheja et al. 2011; Henda 2014)	SPIN	–	Triple DES, Tiger
(Hassan et al. 2018; Hossain et al. 2011)	Debug, RFCOMM	–	SAFER
(Hassan et al. 2018; CYWARE 2019)	SAFER + MITM	–	–
(Mora-Afonso et al. 2013; Haataja 2009; Shrivastava et al. 2012)	Identity based cryptography	–	–

11.5 Summary

Bluetooth enabled communication is very much popular among mobile users. State of the art in Cyber security for Bluetooth shows that there is the urgent need for the safety related to pairing issues while connecting with other mobile devices. Cost is also very critical issue to be taken into consideration for future research works. As per the Table is shown a speed of encryption can be increased with the help of Galois field and advanced encryption standard. As far as key length is concerned there is scope for improvement of the critical period and its security. As per the proposed solution, we can make use of Galois field, various cryptographic algorithms and prime quadratic codes for the purpose of pairing, critical length, and cost. Cost is also one of the tricky things which are considered whenever we buy any application so we can make use of this finite field method to reduce the cost of an application so it will be in the favor of customers who were are going to use that form.

References

CYWARE (2019) malware and Vulnerabilities Blog on Latest Bluetooth hacking techniques expose new attack vectors for hackers, 2019. https://cyware.com/news/latest-bluetooth-hacking-techniques-expose-new-attack-vectors-for-hackers-a16cfb5e

Doon JE (2014) Thieves steal $2 million from gas station ATMs using Bluetooth skimmers. TechWorld News, 23 Jan 2014. http://www.techworld.com/news/security/thieves-steal-2-million-from-gas-station-atms-using-bluetooth-skimmers-3498684/

Haataja K (2009) Security threats and countermeasures bluetooth-enabled systems, Feb 2009

Hassan SS, Bibon SD, Hossain MS, Atiquzzaman M (2018) Security threats in bluetooth technology. Science Direct J Comput Secur 74:308–322

Henda NB (2014) Generic and efficient attacker models in SPIN, SPIN'14, July 21–23, San Jose, USA, 2014

Hossain H, Kabir U, Rahman S (2011) Modified approach to RFCOMM implementation to protect bluetooth technology from Bluebug attack, IJCIT, 2011

Latchmanan S, Parveen S (2018) Applicability of RC4 algorithm in Bluetooth data encryption method for achieving better energy efficiency of mobile devices. In: Proceedings of conference on informatics, Kuala Lumpur, Malaysia. repository.um.edu.my/17597/1/Informatics_003.pdf

Minar NBNI, Tarique M (2012) Bluetooth security threats and solutions: a survey. Int J Distrib Parallel Syst (IJDPS) 3(1):127

Mora-Afonso V, Caballero-Gil P, Molina-Gil J (2013) Strong authentication on smart wireless devices. In: Second international IEEE conference on future generation communication technology (FGCT), 2013, pp 137–142

Patheja PS, Woo AA, Nagwanshi S (2011) A hybrid encryption technique to secure bluetooth communication. In: International conference on computer communication and networks CSI-COMNET-2011

Rajput BS, Gupta P, Yadav S (2012) An integrated encryption scheme used in Bluetooth communication mechanism. Int J Comput Technol Electron Eng (IJCTEE) 1(2):68

Reddy KR, Raju GS (2013) A new design of algorithm for enhancing security in bluetooth communication with triple DES. Int J Sci Res (IJSR), India 2(2)

Sharma T (2016) Attacks on Bluetooth devices. Wegilant Net Solutions Pvt. Ltd

Shrivastava M, Murarka S, Narware S (2012) Performance analysis of quadratic prime code for modified Bluetooth FH communication system. Int J Emerg Technol Adv Eng 2(1)
Vishwakarma P, Patel B (Aug, 2011) Bluetooth security–secure data transfer over Bluetooth. IJCSMS Int J Comput Sci Manag Stud 11(02)

Chapter 12
E-commerce

The enhanced role of Information and Communication Technology (ICT) in everyday human life and the urge of e-commerce applications have given birth to various cyber threats and attacks on them. Sensitive data from 65 organizations all over the world was compromised. This year Verizon 2017 Data Breach Investigations Report includes analysis on 42,068 incidents and 1935 breaches from 84 countries (Verizon 2017). In today's era of Digital world, the use of Internet is very common so as the use of it for online shopping (E-shopping). Nowadays, maximum customers are using online shopping due to its advantages over traditional physical shopping such as low-cost, real-time, interactive, and personalized, cross domain, etc. Due to the use of online shopping, human life has become easier with the good virtual experience. The percentage of customers using the E-shopping is increasing rapidly because it is fast, efficient and which is of economic advantage too. The consumer's interest in the adaptation of E-commerce affected by three factors which are consumer attitude towards online transaction systems, Security and the last is trust in the reliability of online product suppliers. Debit or Credit card fraud and personal information security are major concerns for customers and merchants.

12.1 Introduction

Electronic payments are increasingly becoming part of our everyday lives. For most people, it can be hard to imagine a single day where we do not make a purchase using our payment cards in a physical store or perform some form of online payment or money transfer. Users always take for granted the fact that these systems work, without consideration of how they work. E-commerce and e-business are the ever-growing fields. By 2016, the e-commerce business is predicted to reach $2.2 trillion. The main reasons behind the popularity of e-commerce are the cost and time savings, huge selection opportunities, fuel savings, cost and feature

R. Prasad and V. Rohokale, *Cyber Security: The Lifeline of Information and Communication Technology*, Springer Series in Wireless Technology,
https://doi.org/10.1007/978-3-030-31703-4_12

comparison facilities, etc. E-commerce sales are growing at 20% per year, and it will reach $1.6 trillion in 2015. These statistics shows enormous growth for e-commerce but at the same time, it indirectly warns about the requirement of active cyber security mechanisms to combat against security threats.

E-commerce security is a part of an information security and is applied to the factors such as device security, information security, data integrity, confidentiality, availability, etc. Guarding of E-commerce assets from malign entities and intents like unauthorized access, usage, change and deletion of the important information is nothing but E-commerce security. Service providers, merchants, retailers and customers are the easy targets for the hackers. To increase the percentage of customers, we require a better security mechanism for the E-shopping. Typical E-commerce consists of different processes which are shown in Fig. 12.1.

The organization of the chapter is as follows. Section 12.2 elaborates cyber threats to e-commerce. Security issues in e-commerce are discussed in Sect. 12.3. Section 12.4 highlights security vulnerabilities in e-commerce. Modern security techniques are discussed in Sect. 12.5. Last Sect. 12.6 summarizes the chapter.

Fig. 12.1 E-commerce chain

12.2 Cyber Threats to E-commerce

While dealing with E-commerce, we have to take care of the security issues of online shopping. There are different risks associated with the E-commerce. Finding out system vulnerabilities should go hand in hand with the necessary security provisions which must also assess, evaluate, and resolve the risks involved. Around the globe, the percentage of credit or debit card frauds is around 75% which is an eye-opening thing for e-commerce related people and security developers and researchers as well. Also, the account credit frauds are around 19% which is a massive loss.

Types of online shopping Risks (Li and Ji 2011)

- Judgment risk,
- Information risk,
- Product risk,
- Payment risk,
- Data transmission risk,
- Online-shopping operation risk,
- Distribution risk, and
- After-sale service risk.

E-commerce security has its particular shades and is one of the highest visible security components that affect the end user through their daily payment interaction with business. There are many phases of e-commerce transaction and each stage have different security measures. The tabular form of E-commerce transaction phases and their safety precautions are depicted in Fig. 12.2 (Niranjana Murthy and Chahar 2013).

Generally, the authentication process is dependent on identity management for security provisions and access control techniques to be applied. Conventional

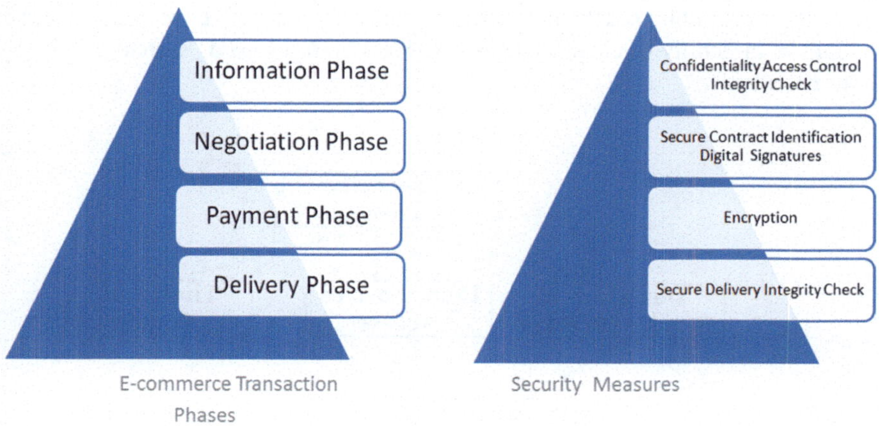

Fig. 12.2 E-commerce transaction phases and necessary security measures

authentication and encryption techniques consume lot of resources which are not suitable for modern E-commerce.

E-commerce transaction phases and necessary security measures are shown in Fig. 12.2. Strong authentication mechanisms consisting of double factor authentication are the need of time now. At each and every e-commerce transaction phase, security measures should be taken care of.

12.2.1 Types of Security Threats

Various security threats such as denial of service, unauthorized access, theft and fraud, spamming, viruses, etc., are depicted in Fig. 12.3. Denial of service attack has been sub-divided into two classes like spamming and viruses. Unauthorized access comes under authentication and authorization issues. Theft and fraud are the other threats to be taken into account while providing security to the system as a whole.

12.2.2 Denial of Services (DOS)

DOS is the type of attack where it removes information altogether and delete information from a transmission or file. The distributed denials of service attacks (DDOS) codes are the common type of attacks which can be implemented within fraction of time. DOS attacks are of two types spamming and viruses.

- Spamming: Spamming consist of flooding the user's mailbox with huge number of mails. DDoS attack can work on damaging mass e-commerce system at the same time.
- Viruses: Viruses are nothing but the specially designed Programs to perform unwanted events. Viruses have the capability to replicate themselves. Viruses are the most publicized threat to client systems. They are effective because of the built-in insecurity of client systems (PC/Mac). Viruses need "system privilege" to be effective.

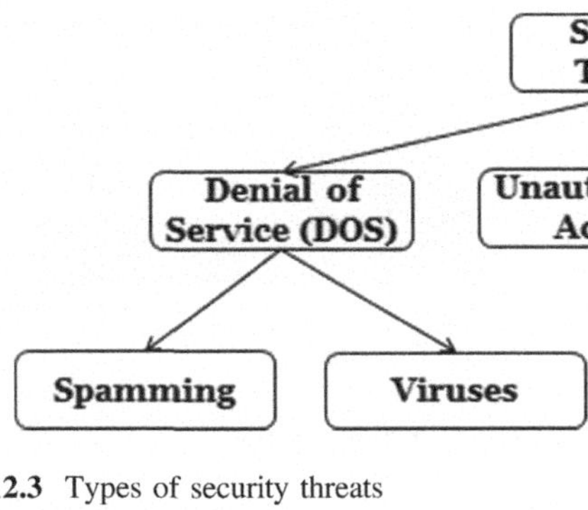

Fig. 12.3 Types of security threats

12.2.3 Unauthorized Access

Illegitimate entrance to the e-commerce business systems, applications or information is termed as unauthorized access. It is classified further into two types. Passive unauthorized access in which just observation of the channel takes place to get the information and that data can be used for launching damage attacks to the system. Active unauthorized access system modifies the system components or the information is changed.

12.2.4 Theft and Fraud

Usage or modification of the stolen data leads to frauds. Thefts and frauds are of various types. It may contain illegal copying of data from some industry's database server or stealing of some useful software. It includes copying credit or debit card details of other users and using them for illegal purchases for selfishness.

12.3 Security Issues in E-commerce

Security is of paramount importance for electronic commerce sites and consumers in the same way. Consumers fear the loss of their financial data and e-commerce sites fear the economic losses associated with break-ins and any resulting bad publicity. Security has become one of the most important issues that must be resolved first to ensure the success of e-commerce. Web Services Security is of vital necessity of the e-commerce world (Farshchi et al. 2011). Three types of the security issues are discussed in the following subsections including client side, server side and transaction side.

12.3.1 Client-Side

Client side security and privacy is the most significant concern from the customer point of view.

In client-side security proper user authentication and authorization, access control, and anti-virus protection is the necessary requirement. For mutual transactions, both customer side and service provider sides should have security and privacy preserving provisions. Communication security is another requirement for secure services. The data analysis on common online banks gives the information about the client side security protection for online banking does need improvement. Two factor authentication is the urgent requirement for online business transactions.

Consumer side safety protection is the weakest part of online banking service providers (IBM et al. 2002). Strong passwords, usable security systems, and self awareness about the possible vulnerabilities and attacks are the much needed strategies for satisfaction of the clients.

12.3.2 Server-Side

The second important issue is server side security issue. It requires proper client authentication and authorization reliability and availability. It should also take care of the non-repudiation of origin, sender anonymity audit trail, and accountability. Availability and denial of service are the most bothering issues for any e-commerce transaction.

12.3.3 Transaction Security Issues

Transaction side security issue is also important to client side and server side security issue. It needs various and better security services such as data authentication, access control, data confidentiality, data integrity and non-repudiation services. Transaction security is critical to bolstering consumer confidence in a particular e-commerce site. There are some defenses for transaction security such as encryption and switched network topologies. Encryption is the most common method of ensuring confidentiality.

12.4 Security Threats to E-commerce

Buying and selling the things by using Internet is termed as e-commerce. Online commercial transactions give relief to consumers and service providers in equal sense. Lot of technologies are supporting e-commerce such as Internet marketing, mobile commerce, electronic funds transfer, supply chain management, inventory management system, automated data collection systems, etc.

With luxury come the threats and attack possibilities. Varieties of threats depending on the application type are depicted in Fig. 12.4. There are threats to online payment systems; there are risks of payment conflicts, e-cash problems, and many more. Online banking has four vertices including financial banking issuers, customers, traders, and regulators (Java T Point Blog 2019). In backdoor attack, hacker gains the unauthorized access to a system and sidesteps the authentication process. This kind of attacker program hides itself from the user and it becomes difficult to sense and eliminate it. Denial of service attack denies services to the legal customers to the important services. The services are made unavailable to the user. In direct attacks, the malicious intent entity gets physical control of the computing device and performs illegal operations on it. It usually contains worms which can steal huge amount of important information from the customer's computing machine.

Eavesdropping is just observation of the happenings across the channels to gain secret information. It does not interfere in the routine activities of the overall system. Credit or debit card attacks are very much sophisticated. The user is told that your credit or debit card validity is about to expire. To renew it, they need number and PIN. Worried user shares the information and within fraction of second, he gets message from his bank about the debit of some considerable amount from his account. Then user comes to know about the white collar robbery.

Data stealing or skimming device is attached to the card reader of the ATM machine of the bank. The information is copied by the reader when consumer swipes the card in the machine. The malign community gets the card details including card number, name, CVV number, expiry date of the card and so on. Phishing is the stylish way of luring the users to click on the fraudulent link sent on their emails or share sensitive credential information by the cyber criminals (Srikanth 2012).

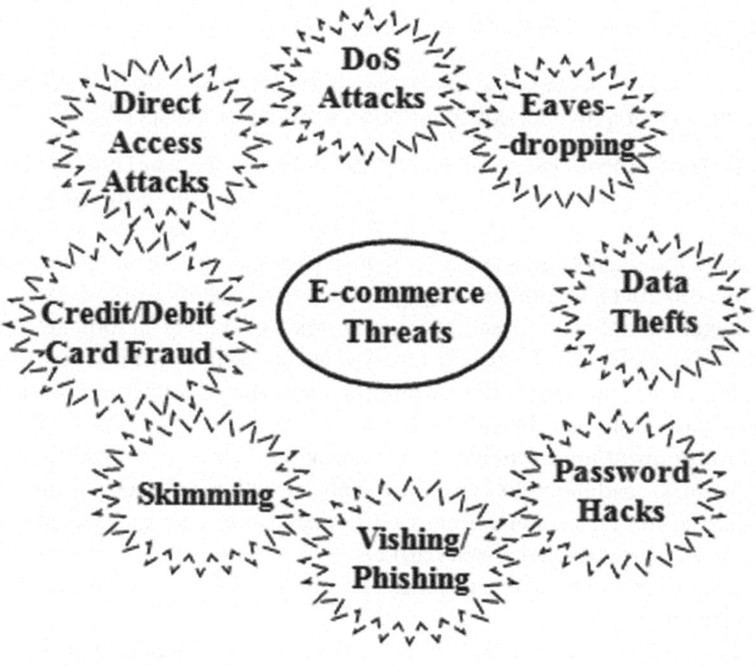

Fig. 12.4 Security threats to E-commerce

12.5 Modern Security Needs for E-commerce

Dimensions of E-commerce security are Integrity, Non-repudiation, Authenticity, Confidentiality, and Availability. The need of time in the forthcoming fifth generation converged mobile communication world is the strong composite cyber security algorithms for every aspect of living. E-commerce is now the integral part of almost everybody's life. Security needs for robust E-commerce are shown in Fig. 12.5. For healthy e-commerce transactions, the security algorithms should be full proof including strong intrusion detection mechanisms which can promptly detect the vulnerabilities present in the system, two-factor authentication that can fail the attempts of masquerades, periodical vulnerability scanning for safety purpose, system integrity monitoring and checking to gain the trust of the customers, regular configuration flaw assessment tests, robust system firewall with the isolated firewall capability for web applications, strong antivirus system and last but not the least appropriate database encryption should be there because most of the frauds happen because the leaked information from such user databases.

Different researchers have proposed the prediction algorithm for the threat prediction and safety issues to E-commerce such as Statistical Modeling and Algorithmic Modeling, out of them; algorithmic modeling is very common technique. But this type of method cannot predict all threats. Newly emerging security mechanisms are data mining and machine learning techniques for threat security (Khan 2019).

Various security techniques which are presently being used include access control, Steganography, and communication security, as explained below.

I. **Access Control**: There are several technologies that can be used to control access to intranet and internet resources. Access control includes authentication, authorization, and audit. It also includes measures of, including biometric scans and metal locks, hidden paths, digital signatures, encryption, social barriers, and monitoring by humans and automated systems.

II. **Steganography**: It describes the process of hiding information within another piece of information (encryption) as well as it Provides a way of hiding an encrypted file within another file. Messages hidden using Steganography are difficult to detect.

III. **Communication Security**: Communications security (COMSEC) is that measures and controls taken to deny unauthorized person information derived from telecommunications. Communications security includes (Yasin and Haseeb 2012; Park et al. 2012):

(a) Crypto security: This includes ensuring message confidentiality and authenticity.

(b) Emission security (EMSEC): Protection was resulting from all measures taken to deny unauthorized person information of value.

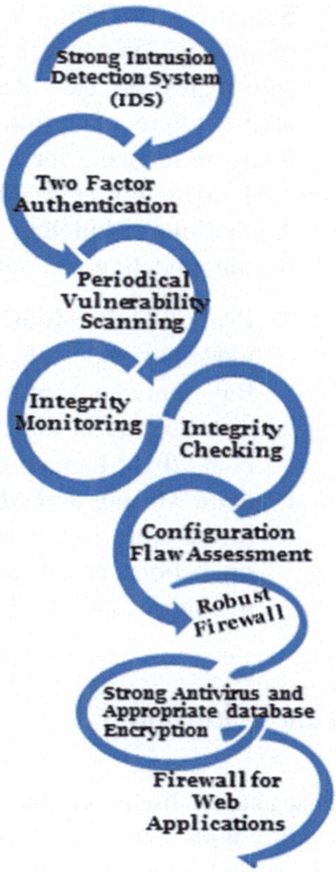

Fig. 12.5 Security needs for robust E-commerce

(c) Physical security: The component of communications security that results from all physical measures necessary to safeguard classified equipment, material, and documents from access there to or observation thereof by unauthorized persons.
(d) Transmission security (TRANSEC): The component of communications security that results from the application of measures designed to protect transmissions from interception and exploitation by means other than cryptanalysis (e.g. frequency hopping and spread spectrum).

The most visible and widely used security technologies are the encryption algorithms. Encryption is achieved by using a mathematically based program and a secret key to produce a string of characters that is unintelligible. The study of encryption is referred as the Cryptography (W3C 2004).

IV. **Security Techniques Using Prediction Algorithm**

For better security in E-commerce, different recent studies are carried out for the predicting cyber-attacks/threats. In the studies, it is found that there different types of threats and to identify them prediction techniques are used such as statistical modeling and algorithmic modeling described in following subsections.

(a) **Statistical Modeling**: It is the most widely used method and employs variants of ordinary least square regression, logistic regression, time-series approaches, auto regression, etc. Classical time-series models use time domain method to predict future value from some combination of the past values, usually with a focus on reducing the systematic error in the model to white noise (Fachkha et al. 2013; Wu et al. 2012).

(b) **Algorithmic Modeling**: It is mainly divided into two categories like probabilistic modeling and data mining and machine learning.

   i. **Probabilistic Modeling**: Out of various technologies belonging to this modeling approach, the most widely used in network security includes Bayesian and Markov Chain method. This method can deal with complex distributions in network traffic, and its results are easy to interpret (Man et al. 2010; Farhadi et al. 2011).

   ii. **Data Mining and Machine Learning**: This method has been widely used in the areas like weather forecasting, earthquake prediction, stock market predictions, etc. It relies on extracting useful information and patterns from the large data sets (Feller 1971).

12.6 Summary

This chapter highlights the existing cyber security E-commerce security threats, security issues and related techniques applied in cyber security field along with the major challenges. Different prediction algorithms are considered for predicting cyber-attacks or threats. Though the prediction techniques being applied in the field of cyber security, but still it requires a lot of research to combat the emergence of new threats in a large computer networks. Usable security and privacy preserving protocols is the need of time for the modern e-commerce assets from unauthorized access, use, alteration, or destruction.

References

Fachkha C, Harb E, Debbabi M (2013) Towards a forecasting model for distributed denial of service activities. In: The proceedings of 12th IEEE international symposium on network computing and applications, Cambridge, MA, pp 110–117

Farhadi H, Amir Haeri M, Khansari M (2011) Alert correlation and prediction using data mining and HMM. ISC Int J Inf Secur 3:77–101

Farshchi SMR, Gharib F, Ziyaee R (2011) Study of security issues on traditional and new generation of E-commerce model. In: International conference on software and computer applications

Feller W (1971) An introduction to probability theory and its applications. Willey, New York

IBM, Microsoft, Verisign (2002) WS-security specification 1.0. http://www.ibm.com/developerworks/library/wssecure

Java T Point Blog (2019) Security threats to E-commerce. https://www.javatpoint.com/security-threat-to-e-commerce

Khan SW (2019) Cyber security issues and challenges in E-commerce. In: Proceedings of 10th international conference on digital strategies for organizational success

Li S, Ji C (2011) Empirical research on perceived risk dimensions of online-shopping infant commodity. In: IEEE international conference on management and service science (MASS)

Man D, Wang Y, Wu Y, Wang W (2010) A combined prediction method for network security situation. In: The proceedings of international conference on computational intelligence and software engineering. IEEE Press, Wuhan, pp 1–4

Niranjana Murthy M, Chahar D (2013) The study of E-commerce security issues and solutions. Int J Adv Res Comput Commun Eng 2(7)

Park H, Jung O, Lee H, In H (2012) Cyber weather forecasting: forecasting unknown internet worms using randomness analysis. In: Gritzalis D, Furnell S, Theoharidou M (eds) Information security and privacy research, AICT, vol 376. Springer, Heidelberg, pp 376–387

Srikanth V (2012) E-commerce online security and trust marks. IJCET 3(2). ISSN 0976-6375

Verizon (2017) Data breach investigations report

W3C (2004) W3C working group note. Web services architecture. http://www.w3c.org/TR/ws-arch

Wu J, Yin L, Guo Y (2012) Cyber-attacks prediction model based on Bayesian networks. In: 18th international conference on parallel and distributed systems. IEEE Press, Singapore, pp 730–731

Yasin S, Haseeb K (2012) Cryptography based E-commerce security: a review. IJCSI 9(2), No 1

Chapter 13
Social Networking Sites

Social networking sites such as Facebook, WhatsApp, Twitter, Instagram, etc. have become the heart of the worldwide wireless network users. These sites are becoming the new weapons for the malicious users. Merely searching with the name of the person, every single detail about that person can be found on the profiles of these sites. This readily available information makes the work of the hackers easier. Cyber world is being compromised by the unauthorized users or hackers. The mainstream news comes in a newspaper how hackers are breaking security for their personal profit or some entertainment. This chapter deals with the cyber security related issues in the usage of social networking sites.

13.1 Introduction

The first worm that affects the world's basic cyber infrastructure which spread over the large area of US is Morris worm. The devil uses all the weaknesses of UNIX system and replicates itself regularly. The Morris worm is developed by Robert Tapan Morris; according to Morris, he was trying to determine the size of Internet. This worm slowed down the computer working speed. He was the first person who was convicted under the US computer fraud and abuse act. Now he is working as a professor at MIT.

More recently, Amazon.com, eBay, Yahoo and some other favorite WWW sites were targets of "denial-of-service" attack. Due to this attack websites suffered about three–four days of massive false traffic. As a result, the sites were in downtime for hours at a time. In January 2011, the Canadian government reported a major cyber-attack against its agencies, including a research organization for Canada's Department of National Defense, Defense Research, and Development Canada. The attack forced the Canada's main economic agencies, to disconnect from the Internet (*Source* Wikipedia).

R. Prasad and V. Rohokale, *Cyber Security: The Lifeline of Information and Communication Technology*, Springer Series in Wireless Technology,
https://doi.org/10.1007/978-3-030-31703-4_13

In fact, a story was in a boom that one of the programmers launched a denial-of-service attack on online stock trading services into his company where he was employed. He breaches the security issue of which the company suffered a lot.

Every organization deployed different security mechanism to protect their data from malicious activities. Some IT companies monitor the incoming and outgoing data traffic to safeguard the business information assets. The rate of expending funds seems to be increasing. However, the companies need to find the mechanism to decrease spending resources for protecting networks from unauthorized access and, employing a method for providing sound security system.

The organization of the chapter is as follows. Section 13.2 elaborates social networking and cyber security. Network security is described in Sect. 13.3. Section 13.4 discusses threat identification. Some strategies for protecting against attacks are explained in Sect. 13.5. Section 13.6 highlights hijacking of social networking sites. Newly emerged security threats to social media are illustrated in Sect. 13.7. Section 13.8 enlists the latest cyber-attacks happened on social media platforms. Last Sect. 13.9 summarizes the chapter.

13.2 Social Networking and Cyber Security

An organization, government agencies, and various corporate companies contain very precise and mass amount of data related to their areas. Following are the reason for which organization need protection and safety. They are:

- To protect company property: This is one of the primary goals of protecting and securing networks. The property means the information which is stored in the systems and sharable by many individuals of the organization. The security in the organization is concerned with protection, integrity, confidentiality and safe access to data.
- For Competitive Advantage: To develop an efficient security system for networks will give the organization a competitive edge. In today's era of Internet financial services and e-commerce, network security assumes prime importance. The customers would avail the services of Internet banking only if the systems are secured.

According to Lohrmann (2010), with the growth of Internet and upcoming social networking applications, the stress on better performance and huge competition has vastly increased. But with these opportunities comes new risk. "Our computing systems cannot just be secure, they should be unfailingly trustworthy," Microsoft founder Bill Gates told World Economic Forum several years ago. "We should be able to rely on them as we in the developed world rely on electricity or a telephone service today."

Cyber Security Breaches are increasing Worldwide, and developing countries must be "Cyber Prepared" to deal with the same. Unfortunately, cybercrime is growing faster than e-government. As of October 1, 2010, privacyrights.org has chronicled 1749 data breaches made public since 2005 that resulted in more than 510 million records being compromised.

Why do we need Cyber Security?

We need cyber security because Cyber criminals are smart beings. They find ways to get into our systems and create havoc in less time than we expect.

- Hackers are everywhere.
- Internet scams and frauds are widespread.
- Cyber theft is a common cybercrime.
- The virus can slow down your computer.
- Spyware, as the name hints, can spy on you.
- Adware can keep unwanted ads to show up.

Social Networks naturally become the part of most literate human being. More than the time they spend on other activities, in a majority they stick with the Facebook, twitter, linked in, etc., the increase in the number of users has amended the hackers to steal or get access to other people's social network profile and use it for some illegal activities. The malicious code injected in the web through different kinds of sources like making the users view eye catching flash ads, redirecting the users to malicious content pages, downloading applications along with malicious code and through the lot many ways the viruses, worms are spread into the web with a purpose of criminal activities involved. The increase in these kinds of spreading malicious codes has become tremendous, and it is done entirely unexposed to the users. Figure 13.1 depicts some of the social networking sites.

Fig. 13.1 Some of the social network sites. *Source* Google images

According to the survey in 2015, Facebook becomes the most widely used SNS. Around 52% of the global population makes use of Facebook. More or less similar will be the percentage for WhatsApp and Twitter and rest of the social networking sites.

13.3 Network Security

Establishing and maintaining a secure network is increasingly very difficult as networks become more and more interconnected and data flows ever more freely. Therefore, it is very important to enable networks to support security services that provide appropriate protection to companies that conduct business and transfer of resources in a relatively open environment. This section explains the breadth of assumptions and challenges to establish and maintain a secure communication networks (SEO 2015). These are the key factors to consider when designing a secure network include the following:

- Business needs
- Risk analysis
- Security policy
- Industry best practices
- Security operations
- Cost analysis.

13.3.1 Basic Security Requirements

To provide appropriate protection to network assets, the procedures, and technologies that you deploy need to guarantee three things, which is also known as CIA triad as shown in Figs. 13.2 and 13.3.

Confidentiality	Providing confidentiality of data guarantees that only authorized users can view sensitive information
Integrity	Only authorized users can change sensitive information and provides a way to detect whether data has been tampered with during transmission; this might also guarantee the authenticity of data
Availability (systems and data)	System and data availability provides uninterrupted access by authorized users to important computing resources and data

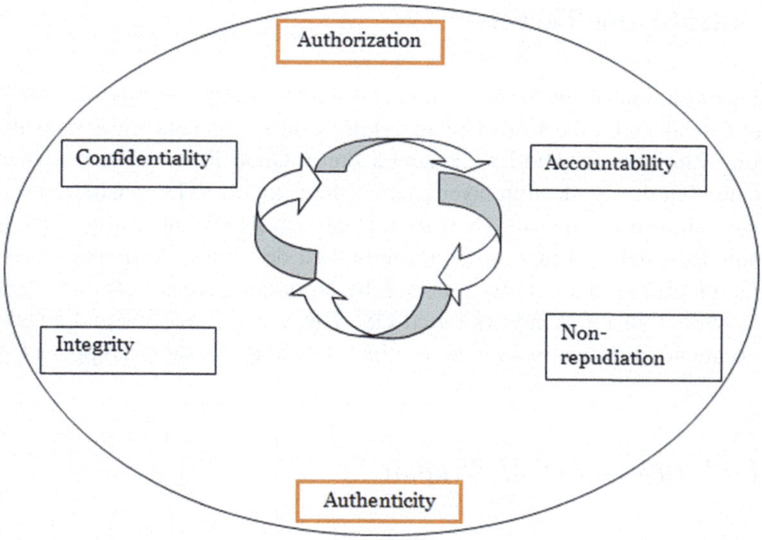

Fig. 13.2 Security parameter wheel

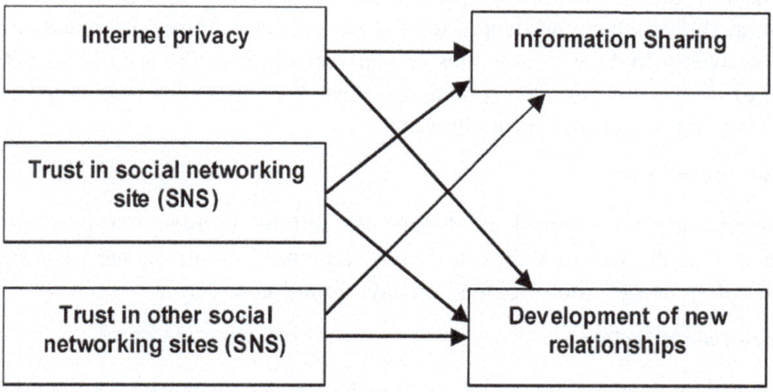

Fig. 13.3 Privacy trust model for SNS

13.3.2 Algorithms Used for the Protection Against Attack

Different approaches and methods have been developed for the protection of network against attacks. Earlier various cryptographic methods are employed for network security, but these methods don't work against attack. National Institute of Standard and Technology (NIST) proposes a method for security uses symmetric key ciphering techniques. Another ciphering known as asymmetric key ciphering are also used for the protection of network security (SEO 2015).

13.4 Identifying Threats

After the Evolution of lot many viruses and worms many systems were exploited and prone to attack by the Virus Creators. For Securing the data while transmission, Mr. Philip Zimmerman who had created a Pretty Good Privacy Protocol which is implemented in the application layer enables the contents to be transferred securely from host to host transmission. Thereafter, understanding the necessity of protecting the computer systems, a private organization had developed Antivirus to disinfect the virus infected systems. Later followed by Norton, Microsoft also developed its own Antivirus tool for Windows based OS. The Viruses, Worms and attacks like DOS are intended to manipulate the original data and misuse (Karuppanan 2012).

13.4.1 Viruses—Lethal Weapon

Malfunctioning of computer, system crash, program hang, theft or damage of data has always been worry for all the system users. Viruses are most lethal weapon against which we have to fight. In order to combat them we have to familiar with them, how it affects and how it spreads. We have to aware how these are being programmed. Let's have an overview of these viruses. Viruses are member of the family malware. Malwares are codes or software to affect the domain of network, data (file) or systems and also spread very fast. We have numerous types of malwares affecting as per their programming.

I. **Boot sector virus**

Files having the boot record are being attacked by viruses. Earlier it came in existence with the use of floppy to boot a computer. With outage of floppy the methods of spread of boot sector virus have found new ways.

II. **Browser hijacker**

Automatic download of files, automatic transfer to web pages which you never wish to open is the most common example of this type of virus. These are generally related to earn revenue through advertisement. As we are aware, every click counts and are being paid. Such type of virus is browser hijacker.

III. **Direct action virus**

Type of a virus which is dormant and start its rampage once it has been executed. These are type of viruses which are old fashioned and rarely exist.

IV. **File infector virus**

Host files when executed starts to work in disruptive fashion, gives us the indication of presence of file infector virus. This type of virus forces the execution infected file instead of execution of indented file.

V. **Polymorphic virus**

Specialty of such viruses is its ability to transform itself with every execution to protect itself from destruction. These viruses keep on altering them as per the condition evolving.

VI. **Resident virus**

Viruses making their home directly in system memory are of this class. This gets executed with the execution of .exe file.

VII. **Web scripting virus**

Malicious code existing in the website content can badly affect the computer after it is being executed. Generally some files are to be executed to support the operation of certain functionality.

13.4.2 Worms

Worms unlike virus do need any support from anywhere for its propagation. Damage done by worms is same to those done by virus, considered as sub class of virus. They are self-sufficient for its execution. It starts with tricking the user and gets to the file transfer system in computer and starts moving unaided. Dangerous property of worm is self-replicating, can create numerous copies without any aid. They consume too much memory space being self-travelling into the network. The gate to worms is being vulnerability in the system.

A computer worm consists of components. They are as follows:

- Target locator
- Infection propagator
- Remote control and update interface
- Life cycle manager
- Payload
- Self-tracking

Different Types of Computer Worms are:

- E-mail Worms
- Instant Messaging Worms
- Internet Worms
- IRC Worms
- File-Sharing Networks Worms.

13.4.3 Web Attack

Web based attacks are one of the greatest and oftentimes the least understood of all risks related to confidentiality, availability, and integrity. The web based attacks focuses on application and layer-7 of the OSI model instead of networks and host. According to John Pescatore of the Gartner group claims that nearly 70% of all attacks occur at the application layer. Application vulnerabilities could provide the means for malicious end users to breach a system's protection mechanisms to take advantage or gain access to private information or system resources.[1]

Over the course of 2008, we observed numerous instances of the following attack techniques:

a. SQL Injection Attacks
b. Malicious Advertisements
c. Search Engine Result Redirection
d. Attacks on the backend virtual hosting companies
e. Vulnerabilities in the Web server or forum hosting software
f. Cross-site scripting (XSS) attacks.

13.4.4 Session Hijacking

The Session Hijacking attack exploits the web session control mechanism, which is normally managed for a session token. This is one of the serious types of attack which enable the attacker to imitate the victim and take over his/her networking session(s). This is because http communication uses different TCP connections. For this the web server needs a method to recognize every user's connections.

13.4.5 Denial of Service (DOS)

In 2000, another one of the serious so-called attack DDOS affects so many high profile web sites to be offline for several hours. This attack was finally identified when the hackers opted for powerful systems from the University of California-Santa-Barbara. This type of attack involves the flooding a computer resources with more requests that it can handle. This causes the resources to crash, denying authorized users the services offered by the resources. Lot of efforts are being put to stop these kinds of viruses and worms but till the day most of efforts are vain. In previous technologies once after detecting the virus it will be analyzed and then counter measure solution will be provided. These kinds of technologies

[1] http://www.ciscopress.com.

will tend to lose the data until a countermeasure solution is provided. To gain the leap forward to protect the system from interferences, analysis at very earlier stage using previously collected data to be done based on the types of attacks and protective measures are to be applied (Al-Hamadi and Chen 2013).

13.5 Protection Against Attacks

For seeking protection against security threats, it is necessary to understand the impact of such attacks and then we can think about the protection techniques.

I. **Impacts of attacks**

- Unwanted disruption or denial of service attacks, including the take down of entire web sites
- Gaining, or attempting to gain, unauthorized access to a computer system or its data
- Automatic installation of viruses without the knowledge of the user
- Unauthorized use for manipulating or stealing the data and many more

II. **How protection can be made against these attacks?**

- Blocking resources which are unnecessary
- Least access to social networking sites

As per the survey data published from web-root it is been identified that 16% of corporate are not at all allowing their employees to use social networking sites like Facebook, Twitter, LinkedIn, etc. 40% of corporate allow partial access to the social networking sites. 46% of corporate allow full access to the networking sites. The anticipation regarding the social engineering attacks is at the risk of increasing. Data's that are prone to be attacked is like Employee contacts, Email Ids, Sensitive information, etc. At Current Scenario to protect against attacks few security solutions followed are

- Firewall
- Endpoint
- Content filtering
- User restriction on device
- UTM server/services on the network

III. **Social engineering attacks through social media**

Traditional Phishing

- Spam
- Banner ads
- Malware and web injections
- Social networking sites

Spear Phishing

- Reconnaissance
- Org chart
- Handpicked targets
- Relevant and convincing mail
- Synchronized attack

IV. **How it can be stopped?**

- Disallow social networking sites
- Educate the Internet users or create awareness about types of attacks and ways to avoid them
- Strong password policy
- Double Authentication
- Use different logins for each service and secure passwords
- Get creative with security questions
- Remove your info from public information databases
- Frequently monitor your accounts and personal data

V. **Multi-layer security**

Multi-layer security is of prime importance because of cross layer functionalities necessary for various tasks in the social media usage. Defense in depth is possible with the help of multi-layer approach. It is expected that it should include anti-spam to block fake social media phishing emails, anti-malware as the last layer of defending to protect against social media malware, and real-time updates delivering automated, round-the-clock protection. Industries also need a simple united platform to manage all of these complexes, 'moving parts' effectively, because a lack of centralized coordinated management and automation can result in security holes. Incorporation of all these things into a single, integrated risk administration platform streamlines the network and improves security protection (Ali et al. 2009).

VI. **Blocking through ACL**

Access control list in the proxy server are meant to control the user's activities and contents from Internet. Apart from that it serves many purposes. So, how this ACL can be used to prevent the social engineering attacks. In the Multi-Layer Security, at some point it requires the data to be analyzed which are experienced. In the ACL, a self-learning system that segregates and simulates the kind of threat classified data can be made possible. The kind of simulation it is performing matters a lot when considered as a protective measure though it consumes few fraction of seconds more. This ACL will work like a self-motivated system that takes care of itself and improves its decision by itself. So, the ACL might function like a neuron activated machine to think itself and develop itself and think like an advanced human being.

13.6 Hijacking of Social Networking Sites

The Session Hijacking attack exploits the web session control mechanism, which is normally managed for a session token.

The session hijack attack contains the following steps:

Step 1 Locating a Target
Step 2 Find an Active Session
Step 3 Perform Sequence Number Prediction
Step 4 Take One of the Parties Offline
Step 5 Take over the Session and Maintain the Connection

The attack also exploits all three sides of the CIA triad as shown in Fig. 13.4. The CIA triad is a representative model used for network security concepts consisting of three basic principles, i.e., Confidentiality, integrity, and availability.

I. **MAN in the middle attack**

The MITM attack is very effective because of the nature of the http protocol and data transfer which are based on ASCII. The man-in-the middle attack intercepts a communication between two systems. The browser creates a SSL connection with the attacker, and by using different techniques the attacker establishes another SSL connection with the web server. Most of the times the browser give warning to the user that the digital certificate used are not valid, but due to the lack of information about threat the user may ignore the warning. In some situation it is possible that the warning does not appear, as for example, when the Server certificate is compromised by the attacker or when the attacker certificate is signed by a trusted CA. Once the TCP connection is intercepted, the attacker acts as a proxy, being able to read, insert and modify the data in the intercepted communication (Asfaw et al. 2010) as depicted in Fig. 13.5.

Another one example is there, in which the attacker capture a session cookie by reading http header, but it is also possible to change an amount of money transaction inside the application context.

Fig. 13.4 CIA triad

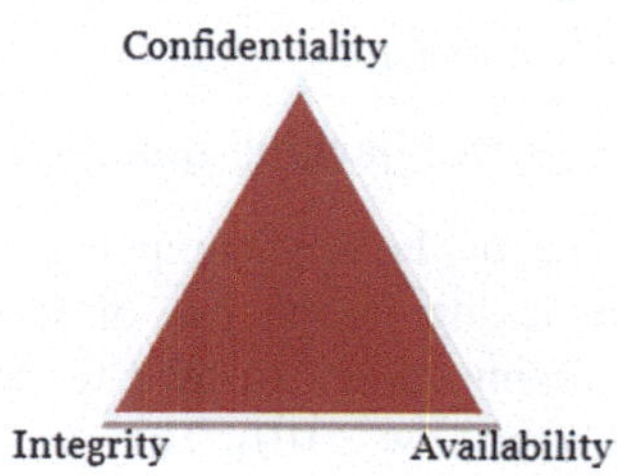

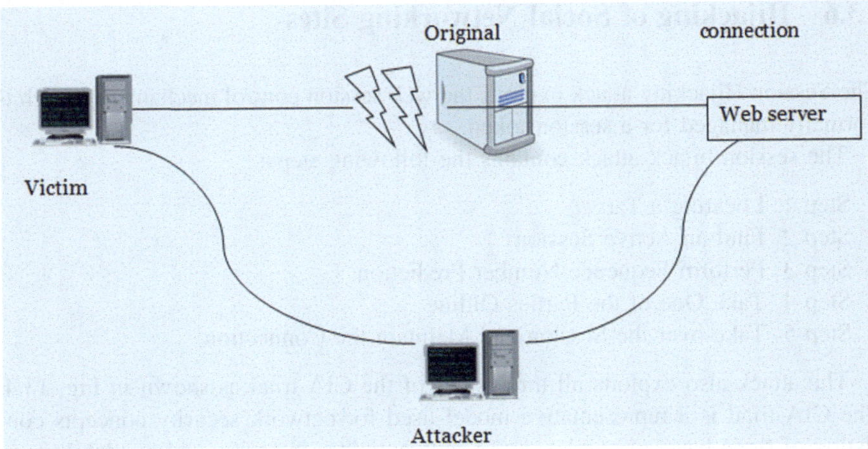

Fig. 13.5 Illustration of man-in-the-middle attack

MITM attack tools

There are several tools to perceive a MITM attack. These tools are particularly efficient in LAN network, because they are capable to implement extra functionalities, like the Arp spoof capabilities that permit the interception of communication between hosts.

- Packet Creator
- Ettercap
- Dsniff
- Cain e Abel

Man-in-the-browser attack

Another type of attack is Man-in-the-Browser attack. This attack uses the same approach as used Man-in-the-middle attack, but in this case a Trojan horse is used to intercept and manipulate calls between the main executable application's browser and its security mechanisms. The intension of attacker behind this attack is to cause financial fraud by manipulating transactions of Internet Banking systems, even when different authentications are in use (Leidigh 2005).

Session hijacking quantum leap

One of the key component for improving security plan or system from session hijacking is the use of Defense in depth. There are several steps a network administrator can take to preemptively protect their network (Asian Schools of Cyber Law 2010). They are:

1. Protect Against Spoofing
2. Use of IPsec and Encryption

3. Use Intrusion Detection Systems and IPS Intrusion Prevention Systems
4. Eliminating Insecure Network Protocols and Operating Systems
5. GPO—Group Policy Objects.

13.7 Social Media Security Threats

In today's technological era, customer's personal and professional life is getting converged on social media platform. Each and every happening is appearing immediately on the social media due to high trust on these platforms. Nobody wish to think about the safety of the data being uploaded on these sites. Every day, new threats are emerging due to excessive use of these sites. Some of these threats are mentioned in Fig. 13.6.

Social engineering has become most famous policy for the cyber attackers. Personal information of the victim user can be easily obtained from their social media accounts. Some important information can be obtained from the target's employee profile and a bogus account can be opened with forged identity of that employee and the attacker tries to build trust. Some extra information is extracted about the industry project and other important things with the help of fake identity thereby dropping a backdoor onto their computing machine.

Fig. 13.6 Security threats to social media

Targeted phishing attacks are generally applied to gain money or steal sensitive information from the user or industry. Fear and anxiety of the target user is used against him to exploit money and/or some important specific information. The success ratio of these attacks is very high because of its specific targeted nature. Bogus accounts with the forged identities is the another threat with social media sites. This fake account owner with forged identity can successfully get connected to thousands of users from various reputed organizations such as government, military, security firms and industry people.

It has become normal practice to open bogus account with the celebrity name. This account is generally used to spread some misinformation and rumors to attract new users which can be later used to implement spamming or phishing attacks. More robust authentication mechanism is the need of time to avoid such kind of threats. The social media site as a whole can be compromised with some malicious code and the visitor to this site can be easily attacked. The advertisements present on the social media sites also contain malicious codes that attracts users to click on them and then hijack the user's computing machine and collect the sensitive information from it.

Spam and malwares can be spread in large amount through social media platforms. It is observed that the cyber criminals often mask their malicious link with the short URL which user think is legitimate and click on it. In a very short time, Botnet can be spread to thousands of computing machines and can infect them. Social media has become a platform where the customers usually share important and sensitive information to the public. The attacker can extract the sensitive information from this and uses it to compromise the system in the industry (Ghosh 2019).

13.8 Recent Social Media Cyber Attacks

The world has experienced number of cyber-attacks related to social media platforms. These are the attacks which are taken place on the social networking sites such as Facebook, WhatsApp, twitter, LinkedIn, etc. The attacks include targeted phishing, malware, data exfiltration, frauds and scams via financial transactions, account hijacking, identity thefts (Wolfe 2017), etc. as enlisted in Table 13.1.

Recommendations for minimizing exposure to social media cyber-attacks (Wolfe 2017):

(1) Interactions with the customers (although trusted) should be limited.
(2) Users should avoid clicking on the links or downloading the file attachments sent through social media.
(3) Two factor authentications is must.
(4) Employees should be trained on things to post and make visible to public.
(5) Automated social media protection tool needs to be adopted by all the stakeholders of the business.

Table 13.1 Recent social media cyber attacks

Sr. no.	Type of attack	Attack description	Year
1	Malware laced posts	Around 10,000 government employees were spear-phished via social media	Jan/ Feb 2017
2	Targeted phishing/ malware	Forged social media identity sends malware to employees. Remote access trojan (RAT) was sent through social media honeypots to get control of victim's computing device	July 2017
3	Twitter counter	Third party mobile application compromised hundreds of high profile accounts. Fake identities were used to hijack user accounts	Mar 2017
4	Malware/data exfiltration	Use of social media by HAMMERS malware	July 2015
5	Fraud and scams via financial offerings	Financial crime through social media	Aug 2016
6	Account hijacking	Social media account of associated press was compromised	Apr 2013
7	Data breach via account takeover	117 million user credentials were exposed by hacking LinkedIn	May 2016
8	Impersonation, fraud and scams, account hijacking	Half million ether coins were stolen by hacking website enigma's slack	Aug 2017
9	Targeted phishing, malware and account takeover	Direct message was sent to customers from compromised brand account via phishing	Sept 2011
10	Targeted phishing and malware	Streaming service Vevo was hacked via targeted LinkedIn phishing attack and 3.12 TB sensitive customer data was exposed	Sept 2017

13.9 Summary

With the increased number of threats to networks such as worms, viruses and DOS, security can no longer be viewed as an option, even within "private" networks. Securing all equipment, resources, and networks these days, is very critical to maintaining uptime and seamless access to services. Defending against the session hijack attack is very difficult because the attack is not dependent on software vulnerabilities, but rather, protocol limitations within the TCP/IP protocol. A variety of methods can be used to reduce your exposure to the attack including intrusion detection and intrusion prevention systems, firewall configuration, IPSec. Session hijacking was discussed which is one of the serious threats to cyber security with their prevention methods. Another problem similar to session hijacking that is session stealing was also discussed.

References

Al-Hamadi H, Chen I-R (2013) Redundancy management of multipath routing for intrusion tolerance in heterogeneous wireless sensor networks. IEEE Trans Netw Serv Manage 19 (2):189–203

Ali KM, Venus W, Al Rababaa MS (2009) The effect of fuzzification on neural networks intrusion detection system. In: Proceedings of the 4th conference on industrial electronics and applications, China, pp 1236–1241

Asfaw B, Bekele D, Eshete B, Villafiorita A, Weldemariam K (2010) Host-based anomaly detection for pervasive medical systems. In: Proceedings of the 5th international conference on risks and security of internet and systems, Montreal, QC, Canada, pp 1–8

Asian Schools of Cyber Law (2010) A to Z of cybercrimes. Department of Science and Technology, Government of India, Lexcode Education and Assessment Platform (LEAP)

Ghosh S (2019) Top seven social media threats. Computer weekly blog. https://www.computerweekly.com/tip/Top-seven-social-media-threats

Karuppanan K (2012) Security, privacy, and trust in social networks. In: Computational social networks: security and privacy. Springer, London

Leidigh C (2005) Fundamental principles of network security. White paper no. 101, American power conversion

Lohrmann D (2010) 5 reasons cyber security should be a top priority

SEO (2015) Archive for the 'social media' category, interview with Mike Ellsworth about social media for B2B, Apr 2015. http://www.seo-writer.com/blog/category/social-media/#ixzz3Yln8mY6E

Wolfe S (2017) The top 10 worst social media cyber-attacks. Blog from info-security magazine. https://www.infosecurity-magazine.com/blogs/top-10-worst-social-media-cyber/

Chapter 14
Secure Incident Handling

When masquerader tries to cheat online user by applying a mixture of tricks for luring the user to share personal information, it results in a phishing attack. Phishing is very crucial kind of attack which usually happens through social sites or frequently visited websites by the user. Incident or event handling is similar to emergency medicine. The caregiver tends to be under pressure, and mistakes can be very costly. After the occurrence of incidence, users have to face other information robberies. Secure Incident Handling is the need of tomorrow's wireless cyber world. This chapter elaborates different cyber security aspects for Incident Handling and Phishing.

14.1 Introduction

Nowadays, the Internet has become one of the most useful and widely available communication mediums on earth, and our life very much depends on it. Governments, banks, schools, and corporations conduct their day-to-day business over the Internet. With such extensive use, the data that resides on and flows across the network varies from banking and securities transactions to proprietary data, medical records, and personal correspondence. The Internet is cheap and easy to access, but the systems attached to it lack a much-needed ease of administration. As a result, many Internet systems are not secure. Additionally, the underlying network protocols that support Internet communication are not secure to use, and few applications make use of the limited security protections that are currently available. The database available on the network makes Internet systems to become more vulnerable attack targets. It is common to see articles in the media referring to Internet intruder activities.

But, however, exploitation of security problems on the Internet is not a new phenomenon. In 1988, the "Internet Worm" incident occurred and resulted in a large percentage of the systems on the network at that time being compromised and

R. Prasad and V. Rohokale, *Cyber Security: The Lifeline of Information and Communication Technology*, Springer Series in Wireless Technology,
https://doi.org/10.1007/978-3-030-31703-4_14

temporarily placed out of service. Shortly after the incident, a meeting was held to identify how to improve response to computer security incidents on the Internet. The recommendations resulting from the meeting included a call for a single point of contact to be conventional for Internet security problems that would act as a trusted clearance house for security information. In response to the recommendations, the computer emergency response team coordination center (CERT/CC) was formed to offer comeback to the computer security occurrences on the Internet. The CERT/CC was one of the first organizations of this type. A computer security incident response team (CSIRT) was formed in 2001 to provide the security response services to any user, company, government agency or organization (Scarfone et al. 2008; West-Brown et al. 2003; Northcutt 2003).

Incident or event handling needs patience and courage. Hurriedly taken steps may result in mistakes which can be very costly. A simple approach is the best. The decent experienced experts follow well defined and systematic steps for responding to security-related incidents. They follow six stages such as preparation, detection, containment, eradication, recovery, and follow-up. They call on others for help (Line et al. 2006). With other incidents, the personal information compromise incidents usually have larger statistics. Chapter is organized as follows. Section 14.2 gives idea about incident handling. How to handle the incident with proper care and security are mentioned in Sect. 14.3. Section 14.4 elaborates activities under information security incident response (ISIRT). Types of incidents are discussed in Sect. 14.5. Lastly, Sect. 14.6 summarizes the chapter.

14.2 Incident Handling

Managing and organizing an effective computer security incident response capability (CSIRC) consist of numerous major decisions and actions. The first consideration should be to form an organization. The organization should decide what type of services the team should provide, consider what kind of team structures can provide those services, and select and implement one or more incident response teams. Incident response plan, policy, and procedure creation are very significant parts of establishing a group so that incident response is accomplished efficiently and meritoriously. The plan, policies and procedures should reflect the team's actions with other teams within the same organization similarly with outside parties, such as the media, and other incident response organizations (Scarfone et al. 2008; West-Brown et al. 2003).

14.2.1 Events and Incidents

Any observable occurrences in a system or network is called event. Events include a user connecting to a file share, a server receiving a request for a Web page, a user sending an email and a firewall blocking a connection attempt. Adverse events are events like a negative consequence, such as network packet floods, system crashes, unauthorized usage of structure, unauthorized access to sensitive data and execution of malicious code that destroys data (West-Brown et al. 2003).

14.2.2 Incident Definition and Examples

Malicious code that is covertly inserted into another program to destroy data which can run destructive programs or otherwise compromise the security, integrity and availability of the victim's data, applications or the whole system is called incident. These codes are designed to perform nefarious functions without the system's user knowledge. Malware consist attacker tools such as rootkits, backdoors, and keystroke loggers, and tracking cookies (The Government of the Hong Kong Special Administrative Region 2012).

An incident may involve any or all of the following:

i. Unauthorized computer access.
ii. Compromise of information integrity.
iii. A denial of service condition.
iv. Loss of information confidentiality.
v. Loss of information availability.
vi. Misuse of systems or information.
vii. Physical damage to systems.

Risk

In general, hurdles are likely to come our way during the completion of the project which effects on its progress. Risk is the "effect of uncertainty on objectives". A Risk is potential problem "it may happen or may not". While the internet is revolutionizing the way business is done, the risk the internet introduces can be fatal to a business. Figure 14.1 depicts different risks during the process in the execution (Alteren 1999).

Virus

A self-replicating code or program that spreads by inserting copies of it into other programs is called a virus. Viruses insert themselves into host programs and propagate when the infected program is executed, generally by opening a file, running a program, clicking on a file attachment. Viruses are designed to play annoying tricks, whereas others have destructive intent (Alteren 1999).

Worm

A worm is nothing but a type of virus that can spread automatically via an e-mail, Internet relay chat or other network transport mechanisms. Worms are completely self-contained. They do not require a host program to infect a victim. They can

Fig. 14.1 Different types of risk

create fully functional copies. They are self-propagating; unlike viruses. Worms take advantage unsecured Windows shares. Although some worms are intended mainly to waste network resources and system, many worms damage systems by performing some malicious acts (Alteren 1999).

14.2.3 Need for Incident Response

The incident response has become an imperative task because attacks frequently cause the loss of personal and business data. Incidents involving viruses, spyware, worms and other malicious code have damaged millions of networks and systems around the world. The following are advantages of having an incident response capability:

i. Responding to incidents systematically so that the appropriate and quick actions can be possible.
ii. Minimizing loss of information and disruption of services.
iii. To provide stronger protection for systems and data.
iv. Appropriate dealing with legal policies and issues that you may face during incidents (Scarfone et al. 2008; West-Brown et al. 2003).

14.2.4 Incident Response Policy, Plan, and Procedure Creation

This section explains policies, plans, and procedures related to incident handling and response, with an emphasis on interactions with outside parties, like law enforcement agencies, media and incident handling or incident reporting organizations. The policy governing incident response is usually strongly individualized to the organization; most of the policies include the same key elements, regardless of whether the organization's incident response capability is indigenous or outsourced.

It is very essential for the organizations to have a formal, coordinated, and focused approach in responding incidents. For effective implementation of such a capability, an organization must have a strong incident response plan. The proposal provides the organization with a roadmap for implementing its incident response capability. The plan should provide a strong approach for how incident response capability can support to the overall association. Each organization needs a plan that meets its unique requirements, which should correlate to the organization's size, mission, functions, and structure. The plan needs management support for effective maintenance and maturity of an incident response capability. The organization's strategies, mission, and goals for incident response should help in determining the structure of its incident response capability.

Procedures need to be based on the incident response plan and policy. Standard operating procedures (SOPs) are a delineation of the checklists, techniques, specific technical processes, and forms, which are used by the incident response team. SOPs should be comprehensive. In addition, while following standardized responses, it should minimize errors, particularly those that might be caused by incident handling tempo and stress. SOPs should be tested to validate their usefulness and accuracy. It should be distributed to all team members. The organization should communicate with outside parties regarding an incident (Scarfone et al. 2008).

14.2.5 Incident Response Team Structure

An incident response team should be available for contact at the emergency situation by anyone who discovers that an event involving the organization has occurred. One or more team members, depending on the seriousness of the incident and availability of manpower, will then handle the incident. The incident handlers first do the analysis like the incident data and then determine the impact of the incident and gives response appropriately to limit the damage to the organization and restore normal services. Even if the team may have very few members, the team's success depends on the participation, motivation, and cooperation of individuals throughout the organization. The incident response team structure identifies such individuals, discusses incident response team models. It provides advice on selecting an appropriate model (West-Brown et al. 2003).

Central Incident Response Team
An incident response team handles instances throughout the organization. This model is effective for both small groups as well as for large organizations with a limited number of computing resources.

Distributed Incident Response Teams
The organization has many incident response teams, each responsible for handling events for a particular segment of the association. This model is effective and efficient for big officialdoms and for the organizations with a large number of computing resources at distant locations. The incident response teams should be part of a single centralized entity because the response process is consistent across the organization and information is generally shared among many people. It is of very much significant because multiple teams may handle similar incidents. Good communication between groups and constant feedback or review makes incident handling more effective and efficient (Scarfone et al. 2008; West-Brown et al. 2003).

Coordinating Team
An incident response team provides suggestion to other teams without having authority like a department wide team may assist individual agencies' teams.

14.3 Handling an Incident Securely

Information security management is a continuous process and iterative cycle. It has several phases, from preliminary preparation through post-incident analysis. The initial phase involves the formation of team and elementary training, and collection of the necessary tools and resources. During preparation, the organization also attempts to limit the number of incidents that can occur by selecting and implementing a set of controls based on the results of risk assessments. The use of proper security protection and safeguard reduces the risk of attacks. Secure incident handling is a continuous process which performs the activities before, during and after some incident occurs. The major phases of the incident response process are preparation, detection and analysis, containment, eradication, recovery and post-incident activity (West-Brown et al. 2003). Figure 14.2 shows different steps in incident response life cycle.

When security incident occurs and the user is unprepared, then follow these basic steps first:

i. Remain calm.
ii. Notify the right people, choose them and get help.
iii. Tell the details of the incident to the minimum number of persons possible for preserving privacy.
iv. Do not use email, chat, talk or news to send some information, instead use phone and fax.

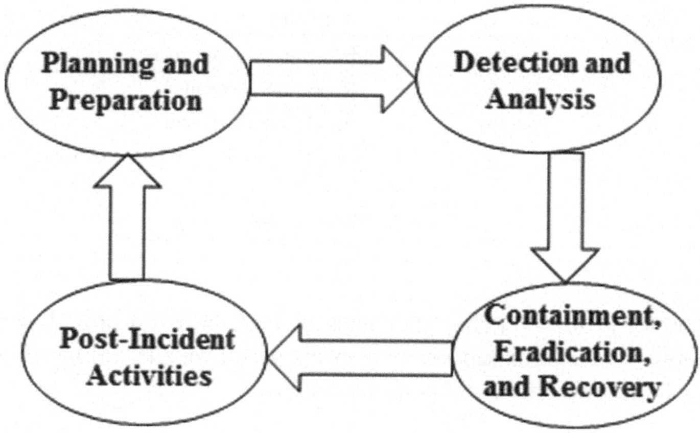

Fig. 14.2 Incident response life cycle

v. Remove computer system from the network.
vi. Prepare a backup of the affected system by using the new medium.
vii. Get back in business by restoring your system from backups to determine whether it can resume its tasks.
viii. Learn from the experience, so you may remain prepared for the next time an incident occurs (Northcutt 2003).

14.3.1 Preparation

Incident response methods typically emphasize preparation so that the organization is ready to respond to incidents. Also, it prevents incidents by ensuring that systems and networks are secure. Although the team is not responsible for incident prevention, it is now considered as a modest component of incident response programs. The incident response team's expertise should be valuable in establishing recommendations for securing systems (Scarfone et al. 2008; West-Brown et al. 2003). Planning and preparation activities are shown in Table 14.1.

14.3.2 Detection and Analysis

Incidents can occur in a number of ways, so it is very problematic to develop a right procedure for handling every happening. So it is important for any organization to be generally prepared to handle common incident types. The incident categories listed below are neither comprehensive nor intended to provide the definitive

Table 14.1 Planning and preparation activities

	Activities in planning and preparation
1.	Incident handling plan
2.	Reporting procedure
3.	Escalation procedure
4.	Security incident response procedure
5.	Training
6.	Monitoring measure

classification for incidents; rather, they merely give a basis for providing advice on how to handle incidents based on their primary category (Scarfone et al. 2008). Table 14.2 depicts detection and analysis activities.

14.3.3 Containment, Eradication, and Recovery

After incident detection, it is very imperative to cover it before the damage increases. Most instances require containment, so it is very crucial to handle each incident. An essential part of containment is decision-making like disconnect it from a network, shut down a system, disconnect its modem cable and disable certain functions. Such decisions are easy to make if strategies and procedures for containing the incident are fixed. Organizations should define acceptable risks in dealing with happenings and develop strategies accordingly. Containment strategies depend on the type of incident. Strong recommendation is that organizations should create separate containment strategies for each type of occasion separately, to facilitate quick and operative decision-making.

In some cases, some organizations delay the containment of an incident so in that case the incident response team should discuss delayed containment to determine if it is feasible. If an organization aware about the fact that a system allows the compromise to continue, it may be responsible if the attacker uses the same system to attack any other systems. The delayed containment strategy is very dangerous because an attacker could escalate unauthorized access or compromise other systems within a fraction of a second. Only a highly experienced incident response

Table 14.2 Detection and analysis activities

	Activities in detection and analysis
1.	The description of the incident
2.	The damage or impact made
3.	Indication if the attacker is still active in the system
4.	Information about the system like organization name, version, functions, host name, IP address, operating system, etc.
5.	Supporting information like screen capture and system messages

team should monitor all of the attacker's actions and disconnect the attacker to attempt this strategy. Containment activities are listed in Table 14.3.

Eradication

The next task after containment is eradication. Eradicating an incident is to remove the cause of the incident from the system, such as removing a virus from the infected data or system. Possible Actions for Incident Eradication are as listed below in Table 14.4.

Recovery

The purpose of this stage is to restore the system to its original state. The recovery stage includes following tasks or activity. Table 14.5 enlists recovery activities.

14.3.4 Post-incident Activity

Post incident activity is one of the most important parts of incident response since it is also the most learning and improving the thing. Each team should discuss new

Table 14.3 Containment activities

	Activities in containment
1.	Conducting impact assessment of the incident on system data
2.	Confirmation if the data or service is actually damaged
3.	Protection of sensitive or critical information and system
4.	A decision on the operation status of the compromised system
5.	Building an image of the compromised system as evidence for subsequent follow-up action
6.	Keeping a record of all actions taken during this stage
7.	Checking any systems associated with the compromised system through any relationship

Table 14.4 Eradication activities

	Activities in eradication
1.	Kill all active processes of the hacker
2.	Delete all fake or image files created by the hacker
3.	Eliminate all malicious programs installed by the hacker
4.	Apply patches and test the system thoroughly before restoring it
5.	Correct improper settings in the system
6.	In the case of a computer virus incident or malicious code, follow the advice of anti-virus tool
7.	In some case make a use of security scanning tools
8.	Update all the passwords of all login accounts
9.	Keep a record of all actions performed

Table 14.5 Recovery activities

	Activities in recovery
1.	Perform damage assessment
2.	Re-install the deleted or damaged files. Sometimes the whole system
3.	Verify that the system is back to its normal operation means the restoring operation was successful or not
4.	Prior notification to all related parties like operators, administrators and senior management
5.	Disable unnecessary services
6.	Keep a record of all actions performed

threats, lessons learned and improved technology. Many organizations have found that holding a meeting with all involved parties after a major incident is extremely helpful in improving the incident handling process itself. This assembly provides a chance to learn by reviewing what occurred, how well intervention worked and what was done to intervene. The convention should be held at the end of the incident (The Government of the Hong Kong Special Administrative Region 2012).

14.4 Information Security Incident Response Team (ISIRT)

An ISIRT is necessary to be established in each bureau/department (B/D). ISIRT is the central body responsible for communication, coordination. It takes security incident handling actions in the B/D. The size and scale of ISIRT varies according to the relative sensitivity of the systems, the measure and scope of the systems in different B/Ds, and potential impact of security incidents on them (The Government of the Hong Kong Special Administrative Region 2012).

14.4.1 Functions of the ISIRT

Following are the main roles of the ISIRT:

i. Overall supervision and coordination of security event handling within the B/D.
ii. Necessary follow-up actions, report to police and further assistance.
iii. Dissemination of security alerts on impending within the B/D.
iv. Information sharing within the B/D on safety incident handling (The Government of the Hong Kong Special Administrative Region 2012).

14.4.2 ISIRT Formation

The ISIRT is coordinating all IT security incidents within the respective B/D. Head of B/D should designate an officer from the senior management team to be the commander of ISIRT. The commander should have the authority to appoint core team members for the ISIRT. In the formation of ISIRT, the advice and support from the Defense Information Technology Services Organization (DITSO) is required to assist the ISIRT commander to develop system-specific security policy and incident handling plan to establish the related logistical arrangements. The DITSO also needs to ensure that the departmental IT security policy is observed and enforced in all the information systems of the respective B/D.

There are some roles that the ISIRT has to play like ISIRT Commander, Incident Response Manager, and Information Coordinator. These duties can be performed by a single officer or may be different captains (The Government of the Hong Kong Special Administrative Region 2012).

14.4.3 Roles of the ISIRT

ISIRT roles vary depending on diverse system entities.

Commander

The responsibilities of the commander are as follows:

i. Decision making on critical matters like system recovery, involvement extent, the engagement of external parties and, service resumption logistics after recovery, etc.
ii. Depending on the impact of the incident on the business operation of the B/D, triggering the departmental disaster recovery procedure where appropriate.
iii. Providing management for the provision of resources for the handling process.
iv. Providing endorsement in respect of the line-to-take for publicity.
v. Coordinating with the Government Information Security *Incident Response* Office (GIRO) on incident reporting and necessary follow-up actions (The Government of the Hong Kong Special Administrative Region 2012).

Incident Response Manager

The role of Incident Response Manager is to monitor all security incidents handling process within the B/D and support for the handling process and seeking management resources. The responsibilities include:

i. Overall management and supervision of security incident handling within the B/D.

ii. Alerting the ISIRT Commander upon receipt of report.

iii. Reporting the status of the security incident handling process to the Commander.

iv. Coordination with various external parties, such as service contractors, to support vendors, Police, and security consultants etc. in handling the incident (The Government of the Hong Kong Special Administrative Region 2012).

Information Coordinator
The role of Information Coordinator is to handle public inquiries regarding the security incident of the B/D. The Information Coordinator is responsible for the overall control and supervision of information dissemination to the public, including the media (The Government of the Hong Kong Special Administrative Region 2012).

Information System Manager
The information system manager will oversee the whole security incident handling process. The responsibilities include:

i. Developing as well as implementing the system specific security incident response procedures.
ii. Observing and following response procedures for reporting incident to the ISIRT of the B/D.
iii. Coordination with all the concerned parties like service providers, contractors, and product support vendors etc. It takes rectification actions against the incident.
iv. Reporting the security incident and requesting for external assistance, such as Police and evidence collection.
v. Providing technical support, evidence collection, system backup and recovery etc. (The Government of the Hong Kong Special Administrative Region 2012).

14.5 Types of Incidents

There are numerous types of incidents like denial of service, malicious code, unauthorized access, intellectual property threats, etc. (BCIT 2005). Table 14.6 depicts the incident types and respective actions to be taken.

14.6 Summary

For an effective information risk management strategy, establishment of appropriate incident response program is of paramount importance. With the continuously growing ICT infrastructure and new mobile applications, the threat environment is rapidly changing and becoming more mysterious. New IT security approaches are

Table 14.6 Types of incidents

Type of incident	Description	Action
Unauthorized access	An individual physical access without permission to system, network, application, data, or other resource	Examine firewall router protections Examine access services regularly
Denial of service	An attack that prevents normal authorized functionality of a system, network, or application by exhausting resources	Employ backups for core services
Malicious code	Successful installation of malicious software like virus, worm, or other code-based malicious entity	Use virus checkers Report suspicious activity Monitor outgoing traffic Protect the software load process Use alternative sources
Scans, probes, attempted access	Any task that seeks to access or identify open ports, protocols, service. This activity does not directly result in denial of service	Report probes to incident response team Assess the damage
Intellectual property	It includes the creative ideas and expressions of the human mind	Inventory your intellectual property Prioritize your intellectual property Assign financial value Implement misuse detection methodologies Stay with the laws
Investigation	Unconfirmed incident which is potentially malicious	Collect maximum data Target analysis (U.S. Department of Justice 2013; Northcutt 2003; United States Government Accountability Office 2014)

very much necessary which can cope up with the increasing threat environments. Creation, provision, and actual operation of incident response capabilities are the need of time. Frequency of attack incidents can be reduced by providing adequate security to networks, systems and applications. Sharing of incident responses with other organizations is also very much essential. Timely detection and analysis of attack incidents is the urgent requirement of ICT. Lessons learned should be used to further improve the security on continuous basis.

References

Alteren B (1999) Implementation and evaluation of the safety element method at four mining sites. Saf Sci 3:231–264

British Columbia Institute of Technology (BCIT) (2005) Industrial security incident database reporting form

Line MB, Albrechtsen E, Jaatun MG, Tøndel IA, Johnsen SO, Longva OH, Wærø I (2008) A structured approach to incident response management in the oil and gas industry. In: Proceedings of the 3rd international workshop on critical information infrastructures security, CRITIS'08

Northcutt S (2003) Computer security incident handling—an action plan for dealing with intrusions, cyber-theft, and other security-related events, version 2.3.1, Mar 2003

Scarfone K, Grance T, Masone K (2008) Computer security incident handling guide. National Institute of Standards and Technology

The Government of the Hong Kong Special Administrative Region (2012) Information security incident handling guidelines, version 5.0, Sept 2012

United States Government Accountability Office (2014) Information security-agencies need to improve cyber incident response practices, report to congressional requesters, Apr 2014

U.S. Department of Justice (2013) Incident response procedures for data breaches, U.S. Department of Justice Instruction 0900.00.01, 6 Aug 2013

West-Brown MJ, Stikvoort D, Kossakowski KP, Killcrece G, Ruefle R, Zajicek M (2003) Handbook for computer security incident response teams (CSIRTs)

Chapter 15
Mobile Device Cyber Security

In 21st century, we all live a fast life as we are surrounded by the technology which made our lives easy and comfortable. A mobile device is a huge invent of technology. In this era, we are totally dependent on mobile devices. Mobile devices have gained lot of popularity because of its portability and high performance. It is a handheld device which occupies everything in it like entertainment, business, shopping and many more. As human beings are continuously using these devices so to make the significant use of these devices is our utmost priority. We store confidential data on it but at the same time security of this device is very important. So in this chapter we will discuss about the major threats and solutions to overcome threats.

15.1 Introduction

There has been evolution in the area of communication and networks. Our daily lives are affected by the devices such as mobile and their use is quite increased. Mobile devices now a day provides a lot of facilities such as social network, banking, online shopping, games etc. Since this device is so useful to us so we need to take care of the security issues related to it. Mobile device security is the full protection of data on portable device and the network connected to the device. Common portable devices within a network include smartphones, tablets and personal computers (Yesilyurt and Yalman 2016).

The chapter is organized as follows. Section 15.2 discusses the capabilities of mobile devices. Various mobile operating systems are explained in Sect. 15.3. Section 15.4 gives idea about security threats to mobile devices. Cyber security measures are illustrated in Sect. 15.5. Section 15.6 summarizes the chapter.

R. Prasad and V. Rohokale, *Cyber Security: The Lifeline of Information and Communication Technology*, Springer Series in Wireless Technology,
https://doi.org/10.1007/978-3-030-31703-4_15

15.2 Capabilities of Mobile Devices

Handheld devices are improvised using advanced tools and technology in order to add new features and thus make it a "featured device". Mobile phone is best example of a featured device. These devices provide more services, capabilities and applications to the users and thus the consumer need is satisfied. Following are the features of mobile phones:-

- Basic function of any mobile device is that it allows us to make calls and send texts from one place to another.
- Every electronic device needs power supply. In mobile phones, battery is the source of power supply to run many applications.
- It is a big source of communication as it provides roaming facility when you are out of your place provided that the two stations have roaming agreement.
- Applications like clock, calendar, alarms, games and music is easily accessible.
- GSM technology is used to make phone calls and send SMS texts.
- GPS provides the feature to track the location of a particular person sitting anywhere. This also provides security.
- A hybrid phone can have more than one SIM card.
- It provides wireless communication so it is physically feasible i.e. we can carry it with us anywhere.
- It is compact and small in size.
- These devices are highly portable because once charged, it can work for hours.
- We can keep all the important data in mobile phones instead of storing them manually. This reduces time and space.
- These devices integrate and connect people globally (Gontovnikas 2018; Raphael 2019).

15.3 Mobile Operating Systems

Operating system is an interface between user and hardware of the system. It makes the communication between user and hardware possible. If the hardware is portable device such as mobile phones, PDA, tablet etc. then it is called as Mobile operating System (Tekade and Shelke 2014). Figure 15.1 shows the different mobile operating system (Mobile OS).

15.3.1 Android

It is an operating system which has open source code that means it is accessible freely or with very little charges. It is developed by Google. Many applications can

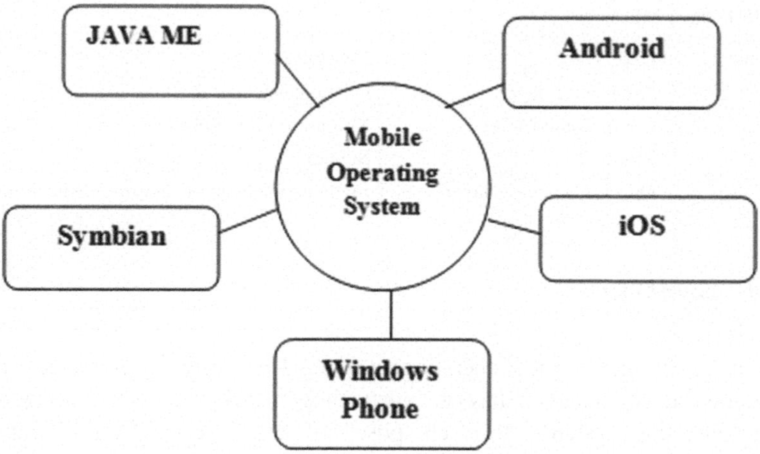

Fig. 15.1 Types of mobile operating systems

be downloaded through Google app store. Android Kernel is based on the Linux. All the personal data, SMS and confidential information can be exposed to other apps (application). Thus it is vulnerable.

15.3.2 Apple Mobile OS

Steve Jobs was the founder of Apple. iOS is an operating system which is restricted to Apple branded products and the applications can be downloaded from the apple store only. It is a closed source operating system. It is based on Darwin operating system. There is no access to external storage. It is not able to achieve the user defined reorganization. Thus it provides more security to the mobile devices as compared to other operating system.

15.3.3 Java ME

Java ME stands for Java Platform Micro Edition. This operating system is designed by Sun Microsystems for mobile and embedded devices. Due to its flexible structure it is widely used. It consists of four layers that are (Table 15.1).

Table 15.1 Java ME layers

Serial no.	Layers	Description
1	Application layer	Contains specific APIs
2	Configuration layer	Minimum class libraries
3	Profile layer	Supports high level services
4	Optional package layer	Involves function or specific applications

15.3.4 Symbian

Symbian is an open source operating system. This software is written in C++ programming language. It has an outstanding security services among all the mobile Operating systems. It is composed of five layers which is shown in Fig. 15.2.

15.3.5 Windows Phone

For smart devices Microsoft developed an operating system known as Windows Phone. Earlier it was developed for PDAs but later on it was used by the smart phones and touch screens. It is a closed source code. All the applications are downloaded using Windows Phone store. The operating system is based on WindowsNT. It has access to external storage.

Fig. 15.2 Layers of symbian

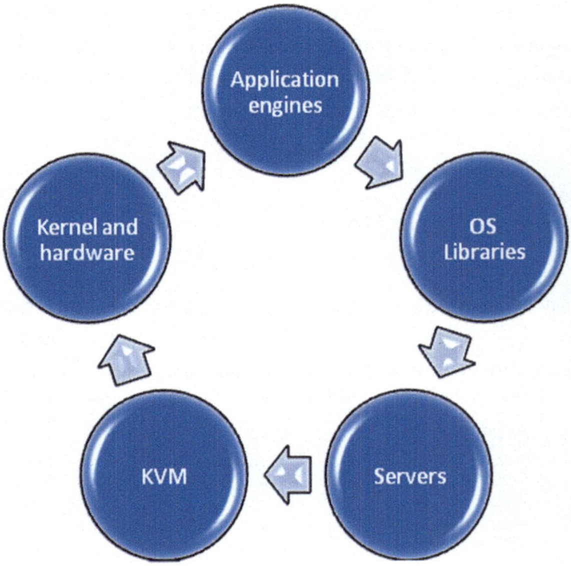

15.4 Mobile Threats

There are lots of threats to mobile devices such as malicious applications, spyware, threats through WiFi. Also threats can be classified based on if they are application based, web based, network based, or if they are physical threats as shown in Fig. 15.3.

15.4.1 Malicious Apps

In today's world people get attracted to the new applications on their mobile devices. While installing any app (application) on your mobile device there is a pop up window every time which ask for permission to few things like media, location, and camera on our devices. These apps can prove hazardous as we grant permission to these apps without actually knowing about them and device becomes vulnerable to these kinds of malicious apps. Due to this our data may lost and damage to confidential storage may take place. So we should look after them with utmost care (Gontovnikas 2018; Tekade and Shelke 2014; Dawson et al. 2019).

15.4.2 Spyware

Irrespective of the operating system the users are having on their mobile phones, their devices are threat targets and are the malign intent actors are looking for

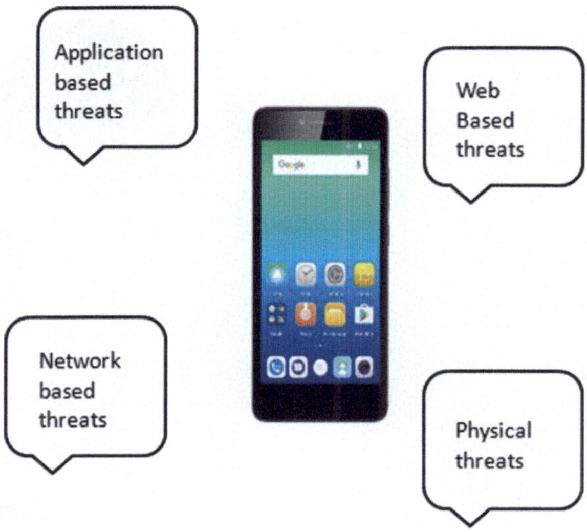

Fig. 15.3 Classification of threats

sensitive information like banking credentials or the industrial data. The big companies like Apple has also experiences zero day vulnerabilities in their mobile devices which kept the devices open for spyware attacks. Recently, Pegasus spyware was found spying on apple devices for surveying users and getting information. Then apple released a patch with updates to protect the users against such kind of vulnerabilities.

15.4.3 WiFi

Now a day's WiFi is used by everyone. Without WiFi it is difficult to work. Public WiFi is required everywhere like schools, offices, airports, railway stations, coffee shops. Using public WiFi is most convenient now days. But these networks are not reliable. We never know because these networks may be operated by the hackers. So our data becomes vulnerable through these devices through networks. Be it iOS or android, both are prone to these kinds of threats (Couture 2010) (Fig. 15.4).

15.4.4 Lack of End-to-End Encryption

It has been found that a lot of data is transferred between different users in a day. It is observed that when using public WiFi, lot of information may get leaked to the hacker. There are various platforms on which employees access corporate data on

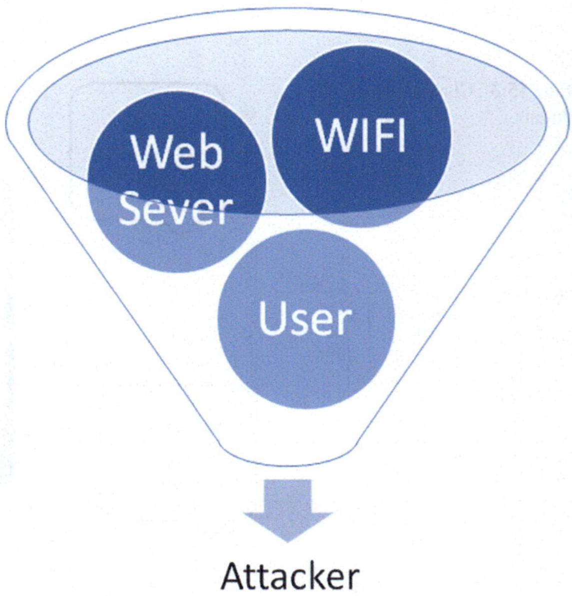

Fig. 15.4 Data leakage to attacker

phones, which results in bad mobile app security. For this purpose, end to end encryption is needed i.e. when data is sent from source to destination then first it should be encoded at source by the process is called encryption. When the data is received at destination then it is decoded by process is called decryption. This is the mechanism of encryption and decryption (Collett 2014).

15.4.5 Inactive Apps

Most of the time mobile devices are having applications which are never used by the users. These inactive apps are becoming the sources of threats. Some of the well known mobile companies remove the inactive apps on regular basis. These inactive apps are the main sources of security threats and privacy breaches. These applications force users to click on the advertisements. Once the user clicks on such kind of ad, it runs in the background without user knowing about it and gathers the sensitive information and generates revenue to the app developer (Gontovnikas 2018).

15.4.6 IoT Mobile Security Threats

IoT which means Internet of things have taken over everything now a day. IOT makes the life of a person convenient with innovative combination of internet and sensor. Now everything is sensed through sensors and sends to the cloud with the help of internet. As the IP addresses are assigned to each network, it becomes easy for the hacker to hack the account of an individual through the respective IP addresses (Beyer 2014).

15.4.7 No Password Protection

Many people even don't lock their devices with the passwords. So if the device is lost or stolen it becomes easy for the thief to access all the data information stored on a particular device. We people are good with hit and trials and hackers are one of us. It is observed that a lot of times people set password in such ways that are easily predictable and hence with the speculations gone right, anyone can hack our accounts. So care should be taken that we should avoid using our information like name, birth date, parents name as passwords because they are easily accessible. Below is the diagram which shows that what should not be used as our passwords (Ruggiero and Foote 2011; Mobile security 2016).

15.4.8 Phishing Attacks

A cybercriminal does a deep search for email addresses of your organization on the internet. They find all publicly available email addresses of your employees. They use these to launch a phishing attack on as many employees as possible. This happens in any organization where employees find that the mail sent to them is legal or authenticated but is actually send by a hacker (Raphael 2019) (Fig. 15.5).

15.4.9 Cryprojacking Attacks

Mobile computing device is hacked, it is shut down without user's knowing and cryptocurrency is asked to make it on again. This type of attack is known as cryprojacking attack. Industry customer devices are used to launch this attack for malign actor's gain. Attack affected phone devices experience poor battery life, slow down operations, or overheating of associated components. Cryprojacking initially started on desktop machines but it emerged as a mobile device threat recently since 2017–2018. There is also the possibility of cryprojacking attack through internet connected set-top boxes which are widely used by customers as well as industries. Hackers are taking advantage of the loopholes present in all Internet connected applications on the mobile devices (Raphael 2019).

15.4.10 Pop up Ads

Adware lures users to click on an ad that directs user to download/install malicious code such as Trojan horse in a word or pdf file. The downloaded may also be the

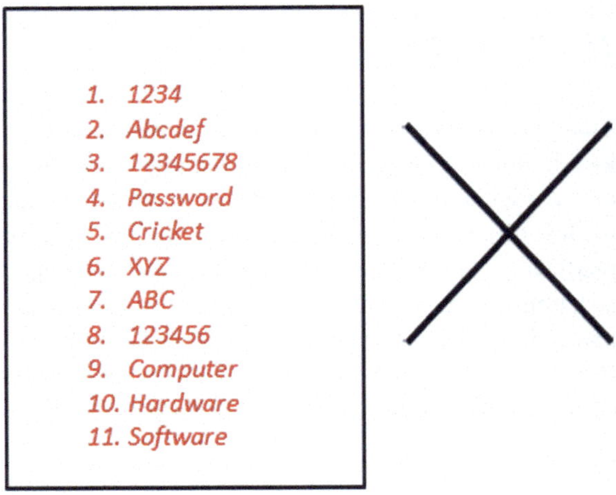

Fig. 15.5 Worst passwords

key logger which monitors the mouse operations or keyboard strokes to steal personal data.

15.4.11 Botnet

One attacker controls a group of sites or devices to send a large volume of traffic to a victim resulted in a denial of service(DoS) attack. Afterwards, the hacker demands the victim a payment to stop the attacking. We can call it a digital blackmailing.

15.4.12 Filtering with Black Listing and White Listing

Many search engines place malicious website a blocked list "blacklist." The search engine will warn to potential visitor who intends access such sites on the list. An enterprise or a personal can also setup their own blacklist. A white list filters only access to these on the list if a white list is exclusive. The filter techniques are widely used for spam email filtering (Zaidi 2016).

15.4.13 Malware Injection

Malware injection is the act of inserting malicious code into a vulnerable web server page with poor application input filtering such that their devices get infected with malware when users interact with such page via form or other GUI components. This injection can be detected by a filter deployed on web server to filter out invalid commands such as SQL injection commands (Zaidi et al. 2016; Selvarani and Ravi 2017).

15.4.14 Lock Bypassing

For security reasons, locks are put on the mobile device apps. The hackers are trying new vectors and techniques for intrusion purpose. The attackers try to bypass these locks by some means and breach the security. Unlatching of the mobile app lock is the technique of defeating the lock that comes under the category of lock picking. If the user has forgotten his Smartphone passcode, then he/she needs the mechanism by which the lock can be bypassed for getting access to the device data. For such kind of ethical lock bypassing, some techniques are there in the practice. But sometimes, with malign intent, to get access to the sensitive information, the lock bypassing is done by the hackers.

15.5 Cyber Security Measures

There is variety of measures for cyber security including password protection, recovery of lost data, malware detection, as discussed in the following subsections.

15.5.1 Password Protection

Password protection is a security process which is used to protect the device to get vulnerable. There are some measures through which private accounts are authorized to authenticated users only. For example, if someone has checked for the mail in other device, at the same time he or she will get the message on his or her device that "Do you just signed in?". There are some protocols for setting the passwords while filling the forms or signing up at any website, Thus in this way password can be protected (More and Kumar 2014; Internet Security Threats Articles 2019).

15.5.2 Recovery of Lost Data

Due to some security threats, data may be lost. So here arises the need of data recovery. In computers we have seen that if in some case some data is lost or deleted, it can be restored back from the recycle bin. In the same way, in mobile devices there is already data backup which is stored on drives so in case data is lost it can be retrieved back. Thus data recovery is defined as the retrieval of lost data from the recovery files or Medias when it cannot be salvaged in a normal way (5 Top Mobile Security Tips 2018; Vylegzhanina et al. 2017) (Fig. 15.6).

15.5.3 Malware Detection

Smart phone has become a must have gadget nowadays. The world has accumulated inside a small mobile device for a user. Hardware and software co-design research has progressed so well that its battery life is improving and size is decreasing day by day. Variety of platforms and operating systems are being used by these mobile devices. Similar to other ICT devices, mobile devices are also prone to malware attacks. Malware detection is the process of scanning the mobile and detecting the malware. This process is quite complex but the detection and removal takes few seconds. There are various techniques of malware detection as discussed in the following subsections and illustrated in Fig. 15.7.

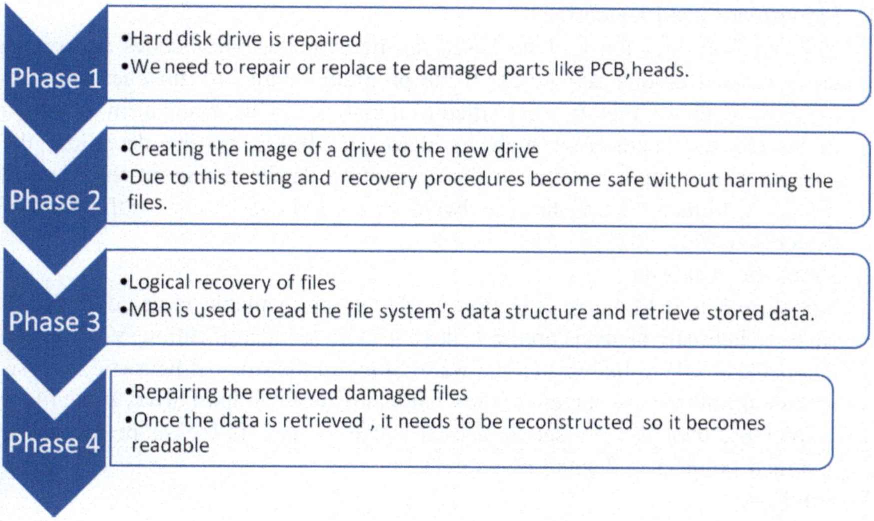

Fig. 15.6 Four phases of data recovery

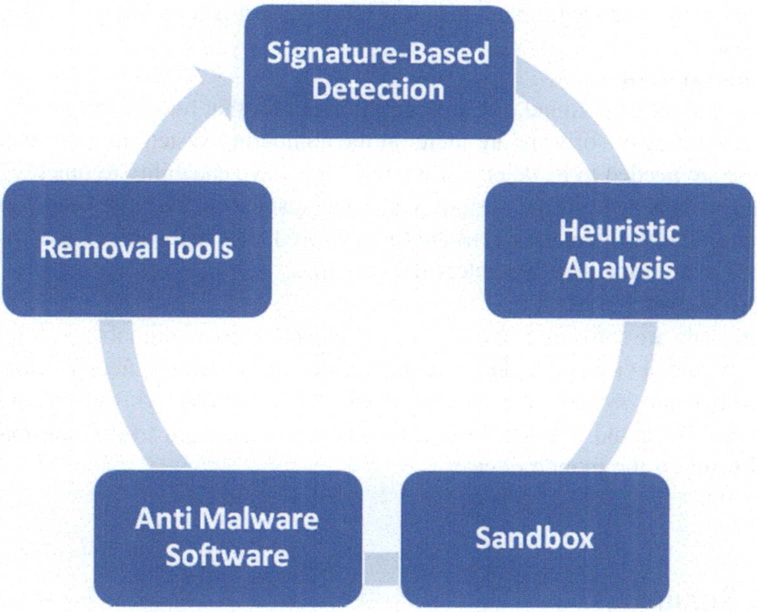

Fig. 15.7 Techniques of malware detection

(1) **Signature-Based Detection**
Malware identification is done based on the available virus codes and previously created unique signatures. If the program's signature matches with the existing signature, then it is classified as a malware. This mechanism is known to be the fast method of malware detection. Based on the detection, the anti-malware system present on the computing device denies of the file and deletes it. Immediate update of malware signature is necessary for this kind of detection (Amro 2017).

(2) **Heuristic Analysis**
Signature Based Malware detection works on the available virus code history whereas heuristic analysis applies rules for malware identification. Violation of these rules leads to listing as a malware. Some of the rules of heuristic analysis include prohibition to camera manipulation, no access to hard drive. If the file is suspicious, then the pre-set numerical value is met. If the score meets the assigned point, it is flagged as a threat.

(3) **Sandbox**
Computing system contains a protected cell created by anti-malware to store suspicious or file from the unknown source is called as sandbox. These files run at other places without infecting other legitimate files. The files are monitored and analyzed inside the sandbox to see if it is harmful. Inside the sandbox, if the file is found to be legitimate, then it is relieved. If it is found to be nasty, then it is denied.

(4) **Removal Tools**
Once the threat is found, the next question comes that how to remove it. If the infected files or software are there on the computing system of mobile device, those are needed to be deleted. Removal tools have capability to quickly delete such malwares from the system. Once the suspicious file is deleted from the computing machine, the tracks are to be cleared. The malware removal process starts every time new file enters the system.

(5) **Anti-Malware Software**
Anti-malware software is used for detection and removal of malware. It protects the mobile device and makes sure that the device is malware free. Without such anti-malware system, the mobile device is vulnerable to various malware attacks. To avoid the data loss, it is very much essential to use anti-malware software on the mobile devices.

15.6 Summary

In this techno-savvy era, smart mobile devices are very much important for us. We are constantly on our handsets to do some useful work and it is equally appreciable that these Smartphone have made our lives fast and convenient. Although the security of these devices is totally in our hands, need to take care that how to protect our data from any illegal activity. We can achieve this by guiding the employees in an office about the measures to be taken to secure the confidential data of the company. Other than employees we can make other people aware about the mobile security threats by campaigning, advertisements or seminars. Technology and security should be considered simultaneously so we can make judicious use of the resources.

References

5 Top Mobile Security Tips to Keep Your Smartphone Safe (2018), 1 Mar 2018. https://medium.com/threat-intel/smartphone-security-tips-f0c30c309030

Amro B (2017) Malware detection techniques for mobile devices. Int J Mob Netw Commun Telematics 7(4/5/6)

Beyer C (2014) Mobile security: a literature review. Int J Comput Appl (0975–8887) 97(8)

Collett S (2014) Five new threats to your mobile device security, 21 May 2014. http://www.csoonline.com/article/2157785/data-protection/five-new-threa

Couture E (2010) Mobile security current threats and emerging protective measures, 2010. The SANS Institute

Dawson M, Wright J, Omar M (2019) University of Charleston. In: Mobile devices: the case for cyber security hardened systems 2019. http://works.bepress.com/maurice_dawson/29/

Gontovnikas M (2018) 10 mobile security threats (and what you can do to fight back), 2018. https://auth0.com/blog/ten-mobile-security-threats-and-what-you-can-do-to-fight-back/

Internet Security Threats Articles (2019) Top 7 mobile security threats: smart phones, tablets, & mobile internet devices—what the future has in store. https://usa.kaspersky.com/resource-center/threats/top-seven-mobile-security-threats-smart-phones-tablets-and-mobile-internet-devices-what-the-future-has-in-store

Mobile security: threats and countermeasures (2016) MKT-6361 V1.0 (2009–2014)

More RM, Kumar A (2014) A study of current scenario of cyber security practices and measures: literature review. Int J Eng Res Gen Sci 2(5). ISSN 2091–2730

Raphael JR (2019) 7 mobile security threats you should take seriously in 2019. https://www.csoonline.com/article/3241727/7-mobile-security-threats-you-should-take-seriously-in-2019.html

Ruggiero P, Foote J (2011) Cyber threats to mobile phones, 2011. Carnegie Mellon University

Selvarani DS, Ravi TN (2017) Issues, solutions and recommendations for mobile device security. Int J Innov Res Technol Sci (IJIRTS). ISSN 2321-1156

Tekade PS, Shelke CJ (2014) A survey on different attacks on mobile devices and its security. Int J Appl Innov Eng Manag (IJAIEM) 3(2). ISSN 2319 – 4847

Vylegzhanina V, Schmidt DC, White J (2017) Gaps and future directions in mobile security research. Vanderbilt University, Nashville, Tennessee, USA

Yesilyurt M, Yalman Y (2016) Security threats on mobile devices and their effects: estimations for the future. Int J Secur Its Appl 10(2):13–26. http://dx.doi.org/10.14257/ijsia.2016.10.2.02

Zaidi SFA, Shah MA, Kamran M, Javaid Q, Sijing (2016) A survey on security for smartphone device. (IJACSA) Int J Adv Comput Sci Appl 7(4)

Chapter 16
Artificial Intelligence and Machine Learning in Cyber Security

Artificial Intelligence (AI) is the intelligence that can be shown by machines, in contrast to the natural way of intelligence displayed by humans or the way in which we can create intelligent machines that work and react like humans. AI is basically a term used when a machine behaves like a human, in activities such as problem-solving or learning, which is also known as Machine Learning. There are so many applications of AI that we use in our day to day lives without even knowing it. Such as, Siri, Alexa, Self-driven cars, Robotics, Gaming etc.

16.1 Introduction

Artificial intelligence techniques are very good at helping us and can also be used to solve cyber security issues. Various techniques come under the umbrella of artificial intelligence such as data mining, neural networks, fuzzy logic that can be integrated with traditional practical and statistical mechanisms for the analysis of data. This data is generally gathered with the help of sensors, filters, exploitation recognition patterns, etc. then the processed data is compared with reference database for supporting the security event management and intrusion prevention (Parati et al. 2008).

Organizations face millions of threats day by day, making it impossible for the security researcher to analyze and figure out the problem, and by the time it responds, it is already too late. But this task can be done by using AI or machine learning in a very effective way. AI allows us to automate the detection of any threat and solve or deal with it even without the involvement of humans leading to securing the data in an efficient way.

Because of its machine driven nature, AI technology assures the users and industry authorities about the error free and state of the art security solutions and services. The benefit of using AI in cyber security is that, these systems can work in a more calculated and in an accurate way resulting in eliminating the errors caused

R. Prasad and V. Rohokale, *Cyber Security: The Lifeline of Information and Communication Technology*, Springer Series in Wireless Technology,
https://doi.org/10.1007/978-3-030-31703-4_16

by humans. Additionally, these systems can do various tasks at the same time, unlike humans. AI is undergoing a vast change and a good progress from a simple technical assistance to helping with cyber security and dealing with the detection and prevention of cyber security challenges.

Artificial Intelligence techniques can bring down any security breach with the help of Machine learning algorithms. The responsible authorities are informed within fraction of second about the events. AI is certainly something of a double-edged sword when it comes to security. While solutions that utilize AI and machine learning can greatly reduce the amount of time needed for threat detection and incident response, the technology can also be used by cybercriminals to increase the efficiency, scalability and success-rate of attacks, drastically altering the threat landscape for companies in the years to come (Brown 2019).

The chapter is organized as follows. Section 16.2 gives idea about machine learning. Different behavioral patterns are deliberated in Sect. 16.3. Section 16.4 throws light on various AI algorithms. Applications of AI in cyber security are illustrated in Sect. 16.5. Section 16.6 discusses AI related available open source tools. Improved cyber security with AI is discussed in Sect. 16.7. Section 16.8 summarizes the chapter.

16.2 Machine Learning

Machine learning is an application of AI that provides systems the ability to automatically learn and improve from experience without being explicitly programmed. Machine learning focuses on the development of computer programs that can access data and use it learn for themselves (Expert system team 2019). The primary aim of machine learning is to allow the computers learn automatically without human interference or assistance and adjust actions and take the decisions accordingly. Machine learning algorithms are often categorized as supervised or unsupervised.

16.2.1 Supervised Machine Learning

Algorithms that can apply what has been learned in the past to new data using categorized examples to predict future events. Initial phase in the supervised ML is the analysis of a known training dataset. After this analysis, inferred function is produced which makes prediction about the outcome. After sufficient training, system can analyze any target and can provide the expected outcome. This kind of learning algorithm can compare the output with the accurate output and can find errors which can help in modifying the model in that direction.

16.2.2 Unsupervised Machine Learning

These types of algorithms are used when the information used to train is neither classified nor labeled. How the system can generate a function which can describe a hidden structure from unlabeled data is studied under unsupervised machine learning. In this case, the system is not able to find out correct output, but by exploring the data, it can draw some conclusions from the available datasets for providing information about the hidden structures from unlabeled data.

16.2.3 Semi-supervised Machine Learning

Algorithms fall somewhere in between supervised and unsupervised learning, since they use both labeled and unlabeled data for training—typically a small amount of labeled data and a large amount of unlabeled data. Learning accuracy can be increased by considerable amount by using this mechanism. When the obtained data needs skilled and significant resources for training or learning purpose, then semi-supervised learning is mostly used (Stevanoic and Pedersen 2016).

16.2.4 Reinforcement Machine Learning

Algorithms is a learning method that interacts with its environment by producing actions and discovers errors or rewards. Predominant characteristics of reinforcement learning are basically trial and error search and delayed reward. Ideal behavior is automatically determined by machines and software agents in this method for the optimum performance. Simple reward feedback is required for the agent to learn which action is best; this is known as the reinforcement signal.

Machine learning enables study of huge quantity of data. While it generally delivers faster, more accurate results in order to identify profitable opportunities or dangerous risks, it may also require additional time and resources to train it properly. Combining machine learning with AI and other technologies can make it even more effective in processing large volumes of information.

16.3 Behavioral Pattern

Emerging AI technologies such as robotics, greater advances in machine learning and high-tech facial recognition form only part of the spectrum of possible innovations. Desire and need to unlock emotional intelligence. Providing greater insight into what human beings are thinking and feeling is the next logical step (Mejia

2019). The ability to exhibit natural behavior patterns (however these may be defined) on a level that is compatible with human behavior is considered a key challenge for robots, virtual agents and intelligent machines designed to interact with humans.

Artificial Emotional Intelligence (AEI) is the category which can provide in depth analysis of the intelligence with emotion and characteristics of the individual human being without any bias influence. Security intelligence can be taken to another level by using the software that can completely remove the human intervention. For real time reading of the subconscious facial expressions accurately and conversion of them into profound emotions, artificial emotional intelligence is a very effective technique.

When used with specific characteristics (real time), it would create an accurate picture. Imagine knowing ahead of time what someone is feeling. The primary emotions the software interprets in these cases are the seven universal languages of emotion; anger, disgust, contempt, happiness, sadness, surprise and fear. In line with this, technology can detect expressions such as condense, passion, honesty, nervousness, curiosity and distress. Knowing when a person is displaying these traits in a public space can be an indicator of potential risk or threat (Tyugu 2011).

AEI technologies can be used for providing effective and reliable security solutions for day to day basis security threats. For real time pattern monitoring and safety provision purpose, AEI can be effectively used. Crowd monitoring and tracking can be done with the facial recognition with the help of AEI algorithms. Machines have started to outperform humans in virtually all areas of rational decision making. Machines can do the right thing. However, when it comes to making bad, irrational or obviously false decisions, most of our behavioral models struggle or fail. Achieving realistic and human-like machine behavior seems to be far more difficult than building programs that outsmart us (Michael 2019).

16.4 AI Algorithms

AI can help a lot in modifying our day to day work and making it simpler. The algorithm is nothing but set of rules that is applied to solve a particular problem. There are some of the AI Algorithms which are used including random forest, k-means, k-nearest neighbors, SVM, Naïve Bayes, decision tree, logistic reasoning, gradient boosting, dimensionality reduction, linear regression, etc. as shown in Fig. 16.1.

16.4.1 K-means

It comes under unsupervised machine learning category that is used for solution of clustering problems. Available datasets are organized into number of clusters

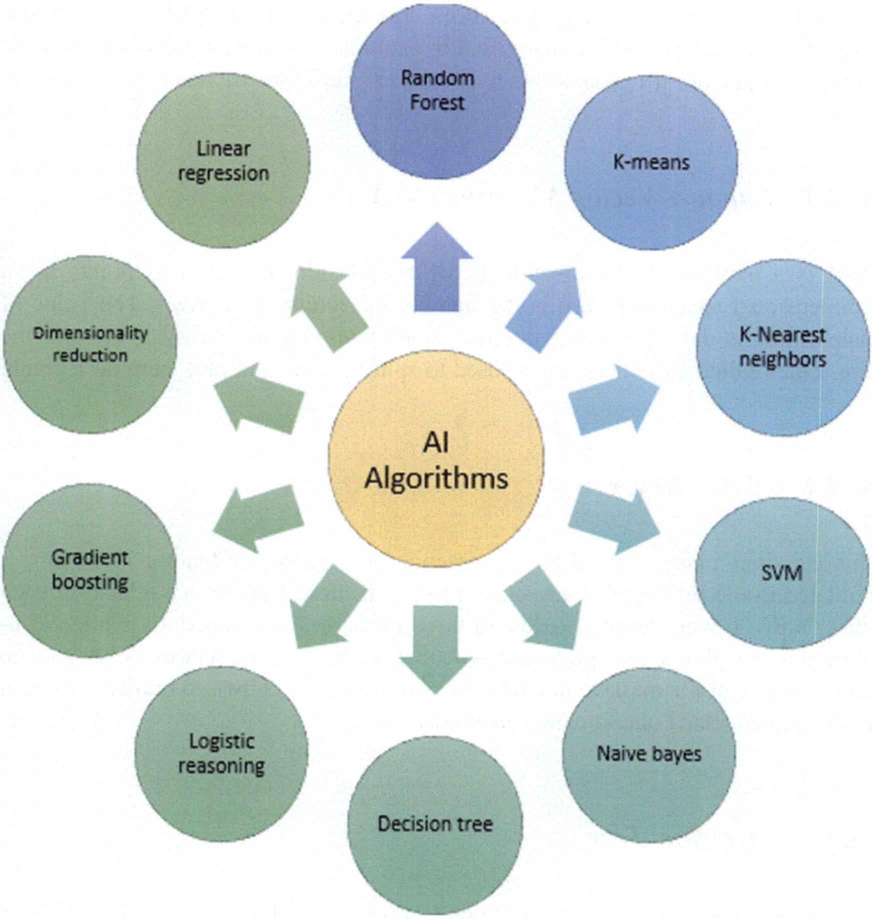

Fig. 16.1 AI Algorithms

depending on its homogeneous or heterogeneous forms. This algorithm prefers centroids which are formed of k number of points known as clusters. It can create new centroids based on the existing cluster members. Closest distance for each data point is determined with the help of these new centroids.

16.4.2 K-nearest Neighbors (KNN)

This algorithm can be applied to both classification and regression problems. It is more widely used to solve classification problems. This algorithm stores all available cases and classifies any new cases by taking a majority vote of its k neighbors. The case is then assigned to the class with which it has the most in common. A distance function performs this measurement. KNN can be easily understood by comparing it to real life. For example, if you want information about a particular person, it makes sense to talk to his family or friends.

16.4.3 Support Vector Machine (SVM)

SVM is a method of classification in which you plot raw data as points in an n-dimensional space (where n is the number of features you have). The value of each feature is then tied to a particular coordinate, making it easy to classify the data. Lines called classifiers can be used to split the data and plot them on a graph.

16.4.4 Naive Bayes

This classifier makes sure that the presence of a particular feature in a class is unrelated to the presence of any other feature. If these features are related to each other, Naïve Bayes classifier treats all these properties independently while calculating the probability of a particular outcome. A Naive Bayesian model is easy to build and useful for massive datasets. It's simple and is known to outperform even highly sophisticated classification methods.

16.4.5 Decision Tree

This is one of the most popular machine learning algorithms in use today, this is a supervised learning algorithm that is used for classifying problems. It works well classifying for both categorical and continuous dependent variables. In this algorithm, we split the population into two or more homogeneous sets based on the most significant attributes/independent variables.

16.4.6 Logistic Reasoning

Logistic Regression is used to estimate discrete values (usually binary values like 0/1) from a set of independent variables. It helps predict the probability of an event by fitting data to a logit function. It is also called logit regression. Logistic regressions models can be improved by the following methods such as include interaction terms, eliminate features, regularize techniques, and use a non-linear model.

16.4.7 Gradient Boosting

These are boosting algorithms used when huge loads of data have to be handled in order to make predictions with high accuracy. Gradient boosting is an ensemble learning algorithm that integrates predictive power of various base estimators for enhancement in the system sturdiness. For building a strong predictor, it integrates weak or average predictors.

Data science competitions like Kaggle, AV Hackathon, and CrowdAnalytix make use of Gradient Boosting techniques.

16.4.8 Dimensionality Reduction

In today's world, massive amounts of data are being stored and analyzed by corporate, research organizations and government agencies. As a data scientist, you know that this raw data contains a lot of information—the challenge is in identifying significant patterns and variables. Dimensionality reduction algorithms like Decision Tree, Factor Analysis, Missing Value Ratio, and Random Forest can help to find relevant details.

16.4.9 Linear Regression

To understand the working functionality of this algorithm, imagine how you would arrange random logs of wood in increasing order of their weight. However, you cannot actually weigh each log. You have to guess its weight just by looking at the height and girth of the log i.e. visual analysis and arrange them using a combination of these visible parameters. This is what linear regression is like. In this process, a relationship is established between independent and dependent variables by fitting them to a line.

16.4.10 Random Tree

Decision trees are collectively termed as random forest. Each tree is classified for classification of new object based on its attributes. The tree has to vote for that class. The classification with the majority of votes is the winner and is selected by forest. If the number of cases in the training set is N, then a sample of N cases is taken at random. This sample will be the training set for growing the tree (Xu 2019).

16.5 Applications of AI in Cyber Security

To detect and avoid cyber security attacks AI is very important. There are many artificial intelligence applications which we use in cyber security solutions. Some of which are shown in Fig. 16.2 such as spam filter applications, fraud detection, botnet detection, secure user authentication, cyber security ratings, hacking incident forecasting, network intrusion detection and prevention, credit scoring and next best offers etc.

Now let us study these applications with reference to AI for cyber security in details.

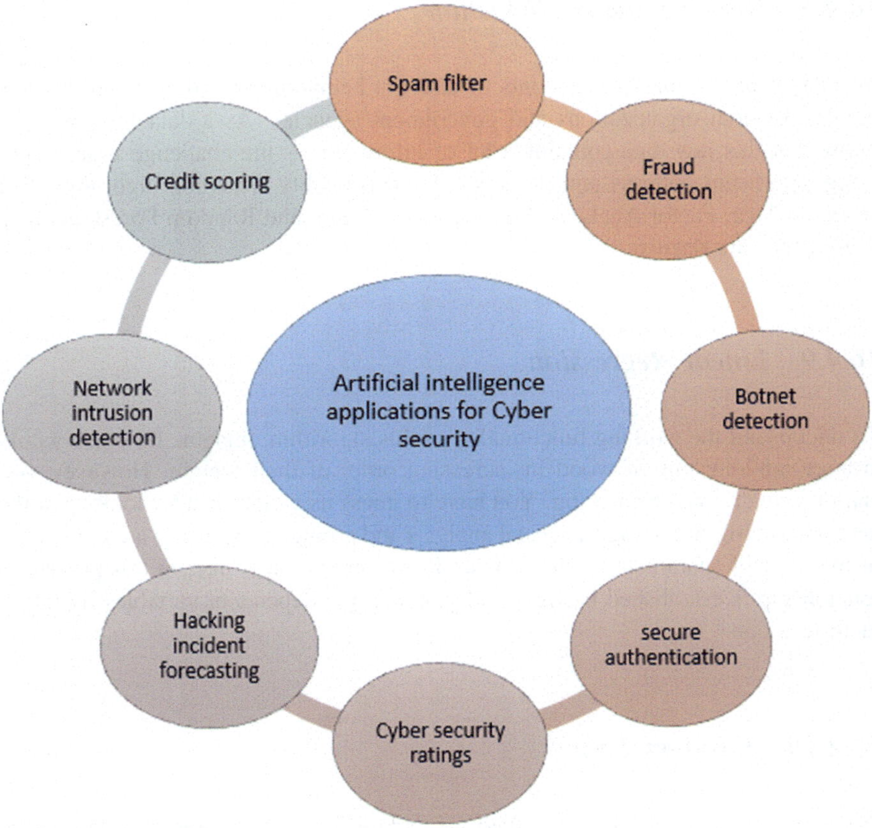

Fig. 16.2 AI applications for cyber security

16.5.1 Spam Filter Applications (Spam Assassin)

A spam filter is a program that is used to detect unwanted e-mails and preventing it from getting into the user's inbox. The spam is kept out of sight of user by using artificial intelligence. The spam filter now uses an artificial neural network to detect and block the spam. Each individual has different kind of mails and hence no inboxes are same. Individual preferences can be reflected with the help of spam filter which is the example of successful machine learning application towards cyber security. Email impersonation, phishing emails can be kept out of the system due to the spam filter technique of machine learning algorithm.

16.5.2 Fraud Detection

Online financial transactions are becoming very popular as well as growing rapidly these days. And so is fraud. Detecting the fraud after the event has happened is of no use. Detection of the suspicious activity and the ability to prevent it from happening is the blessing possible due to AI. This is where artificial intelligence along with the machine learning leads to huge importance. Fraudsters make use of technologies such as machine learning and big data analytics in real time. Fraudsters attack the weakest link.

Fraud detection and prevention becomes very easy through machine learning. Big data analysis is possible with various machine learning techniques and with the analysis, doubtful behavior can be brought to the attention of authorities and those can be cured. Instead of totally relying on humans, the detection and prevention is done by machines. The streaming data can be analyzed on the go and fraudulent signal patterns can be pointed out. It takes a large amount of data to be fed into the machines to get high level of accuracy. Accuracy increases as the machine self-learns and detects the flaws by itself and finds a solution for it.

16.5.3 Botnet Detection

Network of the infected machines is nothing but Botnet. It is used to spread the infection under Distributed denial of service attacks (DDoS) category as well as spamming actions of flooding any inbox or spreading the viruses. Botnet represent network of computers (bots) infected with the same malware and is under the control of hackers (Brown 2019). Through the development of the technology, every personal computer has the great amount of processing power and bandwidth capacity. So every personal computer which is joined into botnet is made botnet more powerful. So, botnet detection and elimination is a very important task in cyber security domain. The accuracy of the botnet detection and elimination depends upon the machine learning algorithm being used for doing so.

16.5.4 User Secure Authentication

Artificial intelligence and its subset such as machine learning and deep learning make it possible to process, verify and authenticate the identity accurately. Machine learning is most useful in identifying whether a customer is real or not. At the time of transaction, artificial intelligence software looks into the person's typical behavior- the way they do their transaction, the devices they use, the way they move the mouse or tap the screen. The software performs the checks and make sure whether the user or the person are the authorized users of the accounts.

16.5.5 Cyber Security Ratings

Rating mechanism is evaluating cyber security infrastructure with some metrics according to some data which are collected passively from the internet. Some of the main purpose of the cyber security ratings is shown in Fig. 16.3. As shown above the purpose of cyber rating are, understanding return of investment, creating a good and compatible cyber security mechanism, support for making tactical decisions, analyze risks and support in VRM and support cyber insurers to see dynamic

Fig. 16.3 Purpose of cyber rating

changes. For rating, systems must work continuously. Continuous scanning of the systems is required to avoid all kinds of cyber-attacks. Machine learning can be used for calculating the cyber security ratings.

16.5.6 Hacking Incident Forecasting

We can predict a hacking incident before it may occur by using AI. In a real world this type of prior predictions can save a lot of money. For doing so, we need a proper dataset which includes latest incidents, reports and some features which can be observed as in externally. The popularity of the company is also one of the major reasons for hacking that particular company's data. It is not always the company's weak cyber security network is what leads to hacking. Passively collected information is used for forecasting mechanism for designing of the cyber security infrastructure. Rating mechanism can be considered as one of the steps in the forecasting, as the main purpose of rating mechanism is evaluating cyber security infrastructure with some metrics according to some data which are collected passively from the internet.

16.5.7 Network Intrusion Detection

Intrusion detection can be defined as the detection of action that does attempts to interfere in confidentiality of the resource. The objective of the network intrusion detection is to identify malicious activities. The most important component which can be used to detect cyber-attacks or malicious activities is the intrusion detection system (IDS). AI plays a very important role here, for detecting intrusions and adapting IDS.

16.5.8 Credit Scoring

Several companies offer AI based credit scoring applications to banks and enterprises creditors looking to better understand the risk associated with their potential borrowers. Traditional methods of credit scoring take into consideration the credit histories of potential borrow. But this might not allow certain people access to credit despite the fact that they could pay their loans back when their payments are due. Ai could allow banks and creditors to score potential borrowers on their creditworthiness using alternative data, specifically which form social media posts and internet activity: what sites someone visits and what they purchase from ecommerce stores. Online behavior can indicate whether a person is likely to pay back their loans. And AI could allow banks and creditors to factor this into their assessments of their potential borrowers (Mejia 2019).

In the present situation of rapidly growing intelligence of malware and cyber-attacks, it is mandatory to develop intelligent cyber defense methods. And AI is one of them which is of crucial importance. It is not clear how rapid development of general artificial intelligence is ahead, but a threat exists that a new level of artificial intelligence may be used by the attackers, as soon as it becomes available (Tyugu 2011).

16.6 AI Related Open Source Tools

For the analysis of AI, large variety of open source and free tools are available. These include Microsoft cognitive toolkit, Theano, Accord.net, TensorFlow, Caffe, Keras, Torch, Scikit-Learn, etc. as depicted in Fig. 16.4 and explained in the following subsections.

16.6.1 Microsoft Cognitive Toolkit

It was released in 2016 and it has proved to be an efficient AI solution to take machine learning work to next level. Deep learning algorithms can be trained to behave like human brains. Some of the features of this toolkit include handling data

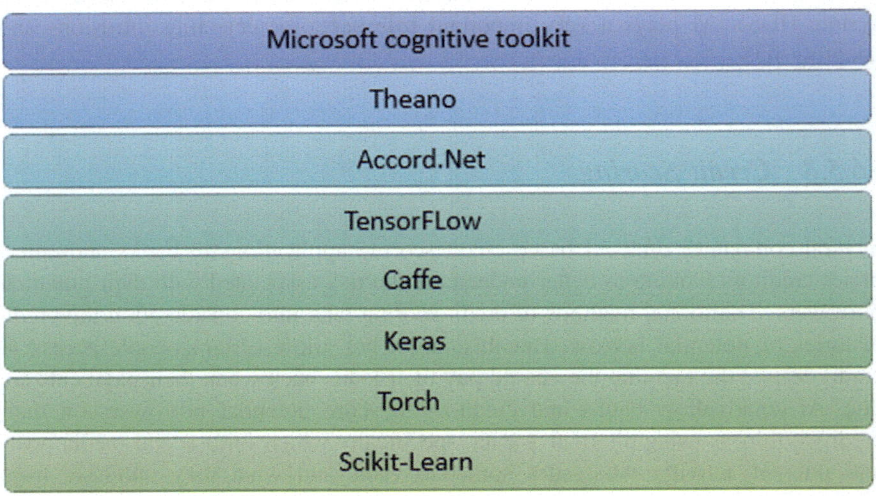

Fig. 16.4 Open source tools for AI

from Python, C++, or Brain Script, ability to provide efficient resource usage, ease of integration with Microsoft Azure, and interoperation with NumPy.

16.6.2 Theano

Variety of machine learning models can be designed easily with the help of Python library present in the Theano tool. It is considered to be industry standard and has motivated innovative developments in deep learning. It helps the user in simplification of the process of defining, optimizing, and assessing mathematical expressions. It accepts structures from user and transforms them into very efficient code that integrates with NumPy, efficient native libraries such as BLAS, and native code (C++).

16.6.3 Accord.Net

Initially released in 2010, Accord.NET is a machine learning framework entirely written in C#. The open source framework is suitable for production-grade scientific computing. With its extensive range of libraries, you can build various applications in artificial neural networks, statistical data processing, image processing, and many others.

16.6.4 TensorFlow

It is an open source machine learning framework that is easy to use and deploy across a variety of platforms and was released in 2015. Initially it was developed by Google and is now used by various big industries such as Dropbox, eBay, Intel, Twitter, and Uber. TensorFlow is available in Python, C++, Haskell, Java, Go, Rust, and now in JavaScript.

16.6.5 Caffe

Initially released in 2017, Caffe (Convolutional Architecture for Fast Feature Embedding) is a machine learning framework that focuses on expressiveness, speed, and modularity. The open source framework is written in C++ and comes with a Python interface. Caffe's main features include an expressive architecture that inspires innovation, extensive code that facilitates active development, fast performance that accelerates industry deployment, and a vibrant community that stimulates growth.

16.6.6 Keras

Initially released in 2015, Keras is an open source software library designed to simplify the creation of deep learning models. It is written in Python and can be deployed on top of other AI technologies such as TensorFlow, Microsoft Cognitive Toolkit (CNTK), and Theano. Keras is known for its user-friendliness, modularity, and ease of extensibility. It is suitable if you need a machine learning library that allows for easy and fast prototyping, supports both convolutional and recurrent networks, and runs optimally on both central processing units (CPUs) and graphics processing units (GPUs).

16.6.7 Torch

It was released in 2002 and it is basically a machine learning library which supports various deep learning algorithms. Torch's key features include N-dimensional arrays, linear algebra routines, numeric optimization routines, efficient GPU support, and support for iOS and Android platforms.

16.6.8 Scikit-Learn

Scikit-learn is an open source library developed for machine learning and it was released in 2007. It is written in Python and features several machine learning models including classification, regression, clustering, and dimensionality reduction. Scikit-learn is designed on other three open source projects such as Matplotlib, NumPy, and SciPy thatare focused on data mining and data analysis (Michael 2019).

16.7 Improved Cyber Security with AI

Cyber security has improved a way lot due to AI. AI has ensured to a very good extent that less cyber related crimes are taking place. AI helps us to secure cyber environment in the ways shown in Fig. 16.5 and explained in the following subsections.

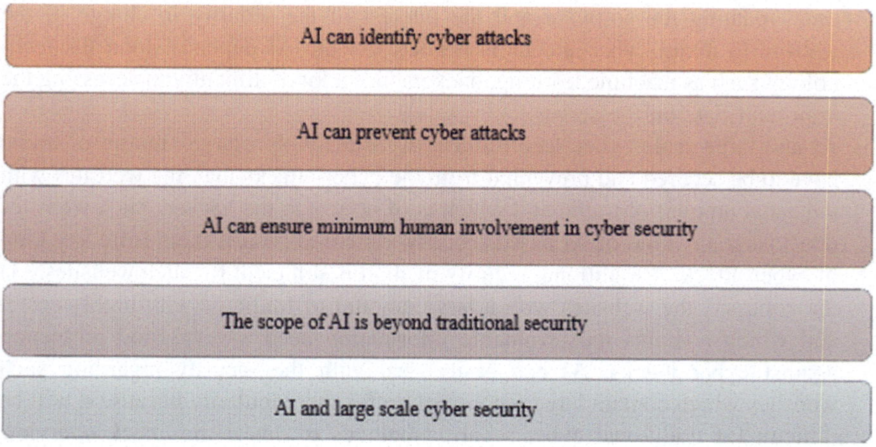

Fig. 16.5 Improved cyber security with AI

(1) Artificial intelligence can identify cyber-attacks: AI can be used to identify cyber-attacks prior to any attack that may happen unexpectedly. Hackers present worldwide use different methods to hack and initiate the cyber-attack. So the organizations or sites which need high security totally depend upon AI to provide high security. And hence, it becomes difficult for hackers to access the sites or any important information because these use AI for their security. As AI can identify the cyber-attacks before it may even happen makes it a good security provider against these kinds of attacks made by hackers.
(2) Artificial intelligence can prevent cyber-attacks: only the identification of the threat is not sufficient and cannot actually help the sites to get rid of hackers and the cyber-attacks. AI can be used to prevent the cyber-attacks in many ways. Here the AI has to think in the way the hacker can think of hacking the website and then act accordingly. Hackers tend to keep an eye on the website they have to hack and find the ways and procedure to do so.
(3) AI can ensure minimum involvement of humans in cyber security: If AI not used, the security heads have to keep an eye on the websites continuously and hence they have to keep the control all over the website by taking the security related decisions. But it is practically impossible to do so. It is impossible for the security heads to continuously monitor the websites without any breaks and interruption. Hence, AI comes to the rescue. AI can continuously monitor the website as well as take the required decisions regarding the cyber-attacks accordingly. AI can help in reducing and eliminating the human errors that could have been caused if AI was not used.
(4) The scope of AI is beyond traditional security: Traditional security methods contain certain antivirus software, firewall and the tools used for detection and prevention of web based security threats. By seeing this scenario, the time to

time updating the software and the interest of the security in charge of the website to avoid cyber-attacks is needed. As the AI depends upon the technology such as machine learning, hackers face a lot of difficulty in accessing the important data on the server.

(5) AI and large scale cyber security: The websites with a large amount of traffic have to be secured and prevented from the cyber-attacks than the websites with the usual amount of traffic and which is of no use to the hacker. Such websites don't need any kind of AI based security system to protect them from any kind of cyber-attack. A traditional security method is sufficient for such websites. On the contrary, the websites with a large amount of traffic or with highly secret and which contains most valuable information need a very good protection against cyber-attacks. AI can easily deal with the security regarding such websites which contain large amount of traffic and popularity because it will be difficult for traditional cyber security methods to detect the attack overview (Brown 2019).

Hence, we can say that AI is undergoing a vast change and a good progress from a simple technical assistance to helping with cyber security and dealing with the detection and prevention of cyber security challenges.

16.8 Summary

AI is adding value to the security sectors of the corporations and individual persons as well, it has helped a lot in eliminating cyber-attacks to a great extent resulting in securing the data for companies and avoiding many frauds that could happen. With the added benefit of machine learning, AI algorithms can easily turn down the cyber security breaches and with the proper information, the security personnel can rectify the system within fraction of time. So, it is evident that the artificial intelligence plays a very important role in cyber security. There is huge scope for upcoming researchers to work in this field.

References

Brown T (2019) IT chronicles, How AI is changing the cybersecurity landscape. https://www.itchronicles.com/security/how-ai-is-changing-the-cybersecurity-landscape/

Expert system team (2019) Expert system, What is machine learning? A definition. https://www.expertsystem.com/machine-learning-definition/

Garbade MJ (2019) Top 8 open source AI technologies in machine learning, opensource.com. https://opensource.com/article/18/5/top-8-open-source-ai-technologies-machine-learning

Mejia N (2019) EMERJ, AI for credit scoring-An overview of start-ups and innovation. https://emerj.com/ai-sector-overviews/ai-for-credit-scoring-an-overview-of-startups-and-innovation/

Parati N, Malik L, Joshi AG (2008) Artificial intelligent based threat prevention and sensing engine: architecture and design issues. IEEE 978-0-7695-3267-7/08

Stevanoic M, Pedersen JM (2016) On the use of machine learning on identifying botnet network traffic. Aalborg University, Denmark

Tyugu E (2011) Artificial intelligence in cyber defense. In: Czosseck C, Tyugu E, Wingfield T (eds) 2011 3rd international conference on cyber conflict. Tallinn, Estonia

Xu Y (2019) Artificial emotional intelligence: understanding human behaviour, Blog comparethe-cloud. https://www.comparethecloud.net/articles/artificial-intelligence-human-behaviour/

Chapter 17
Blockchain Technology

More than hundred countries have experienced WannaCry Ransomware cyber-attack in 2017. This attack demanded around $300 of Bitcoin cryptocurrency from users to retrieve their data. There onwards, Blockchain technology and cryptocurrency are catching the attention of everyone. Blockchain is a growing digital record keeping register in the distributed form which cannot be altered. In case of Blockchain, no central authority is needed. Modification of the data or assets stored on Blockchain is highly difficult or nearly impossible. All the data available on Blockchain can be publicly seen by everyone. It cuts down the excessive costs required in the conventional manual record keeping registers. Blockchain technology has lot of bonus points including integrity, write access, reduced cost, trust, etc. With Blockchain technology, various tokens are available such as currency tokens, asset tokens, utility tokens, and equity tokens.

17.1 Introduction

A peer-to-peer electronic cash system that is Bitcoin was introduced by Satoshi Nakamoto in 2008. For online payment of electronic cash, peer-to-peer transaction system was proposed by the author that did not require any financial body in between. Newly created transaction blocks are added to the existing chain of hash based and time stamped blocks which form an ever growing chain of blocks. In this system, the messages are broadcasted to all the nodes and on individual basis and as per their wish, nodes can enter or leave the Blockchain network without having chance to modify or delete the existing contents. The individual node make use of their CPU power for voting to accept legitimate blocks and to decline illogical blocks by turning down the request of working for such illegitimate blocks (Nakamoto 2008).

Blockchain is a reward driven system for obtaining agreement among all the network entities. Consensus mechanism is the strength of Blockchain technology.

R. Prasad and V. Rohokale, *Cyber Security: The Lifeline of Information and Communication Technology*, Springer Series in Wireless Technology,
https://doi.org/10.1007/978-3-030-31703-4_17

The conventional transaction systems need a trusted third party authority or central server. Due to the presence of third party in the transaction, the double spending problem has emerged. Blockchain for Bitcoin invention proved to be the first digital currency to avoid third party intervention in between two transacting entities. A list of digital record is prepared with the help of cryptography in such a way that every block is a cryptographic hash of the previous block, timestamp, and transaction payload data. Once a block becomes a part of Blockchain digital ledger, it is nearly impossible to make changes into the block transaction data. Due to this property of immunity to data modification, Blockchain is known to be almost incorruptible.

The chapter is organized as follows. Section 17.2 describes the working principle of Blockchain Technology. Blockchain systems and various cryptocurrencies are illustrated in Sect. 17.3. Section 17.4 depicts some applications of Blockchain. Various threats and attacks are discussed in Sect. 17.5. Section 17.6 discusses about cyber security revolution with Blockchain. The chapter is summarized in Sect. 17.7.

17.2 Blockchain Technology

Blockchain is a great technical innovation which can shift businesses from centralized way to decentralized natured trustworthy system. Blockchain has the capability to change the ways of traditional transaction types and can provide different aspects to other parts including multi-party computation, decentralized autonomous institutes, and for other government applications (Jesus et al. 2018). Next subsections elaborate the evolution of the Blockchain and its working principle respectively.

17.2.1 Evolution of the Blockchain

The pace of Blockchain technology is inevitable. It has evolved from Blockchain 1.0 to Blockchain 5.0 gracefully with additional functionalities at each growing step. Blockchain 1.0 is a commercial currency application that is cryptocurrencies. This version mainly works with money transfer, payment mechanism, and digital payment systems influenced by the use of Bitcoin and its variants. Blockchain 2.0 is related with smart contracts. Blockchain makes the smart contracts so robust that it is nearly impossible to corrupt these contracts. These smart contracts include the complete list of financial matters, market, economic applications, stocks database, share market documents, various bonds, loans, advances, and other simple contracts. Ethereum Blockchain allows the execution of smart contracts (Unibright Blog 2017). Blockchain 3.0 is dedicated for variety of decentralized applications in the areas of health, science, and management. This version aims at uses ahead of money, economics, and markets. The evolution of Blockchain with all these

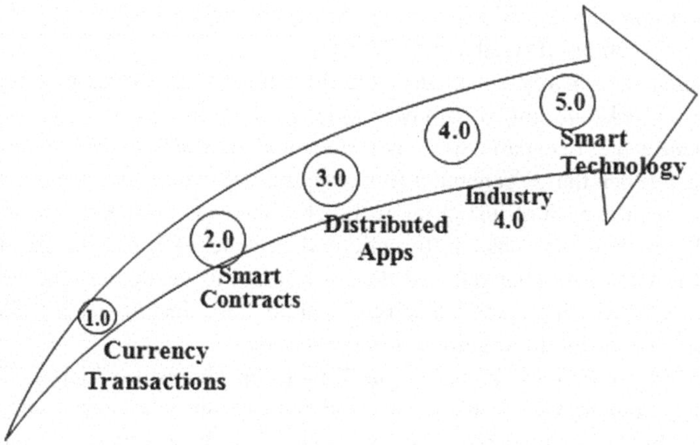

Fig. 17.1 Blockchain evolution

versions is depicted in Fig. 17.1. Blockchain 4.0 endeavors to make Blockchain utilizable for Industry 4.0 demands. Industry 4.0 is meant for automation, industrial resource planning, and incorporation of various other technological systems. For improved version of trust and privacy defense, Blockchain has to play very important role. Some of the applications include supply chain management, approval workflows, financial transactions and condition based payments, IoT data collection, health management, asset management, banking, real estate, entertainment, tourism, games, social media, and collection of information, etc.

Blockchain 5.0 is looking forward to serve for exclusive smart technologies. Improved version of Blockchain 5.0 needs to be more innovative with great speed of operation. Variety of multi-national industries like Alibaba, Apple, Google, Samsung, and Microsoft have kick started their efforts in the direction of research and development for this upcoming breakthrough technology (Pro Blockchain Media Blog 2018).

17.2.2 Blockchain Working

Digital currency Bitcoin is the most prominent application of the Blockchain technology. Bitcoin is the electronic conception and it does not need any central authority to approve the financial operations. Bitcoin is an electronic cash system which performs the economic transactions in a safe, certifiable, and rigid approach. Decentralization is the main virtue of Blockchain technology by which third party interventions during transactions are completely bypassed. Integrity of the transactions is maintained with the help of cryptographic protocols and algorithms. Bitcoin transactions do not need the identity of the sender. It maintains pseudonymity and limited supply characteristics of Blockchain. Immutability and divisibility properties of Blockchain make it unique technological solution for variety of transactions (Priyadarshini 2019).

Blockchain is basically a data structure that stores transactions in a block form and all these blocks are linked with each other in chain like manner. The individual block structure has two parts such as header and transaction data. Unique ID is given to each block that is generated from cryptographic process. Header contains a field that stores the hash of just previous block. Since each block is connected with previous block with the hash information, if anyone tries to change any block contents, the hash gets changed and then it is must to recalculate the hash of all previous blocks which is practically not feasible. This feature makes Blockchain more secure for all the transactions it is containing.

Blockchain technology is based on Bitcoin mechanism which is based on consensus system to which all the network entities need to agree on. Working principle behind cryptocurrency transaction in a Blockchain is depicted in Fig. 17.2. If a user wishes to put some contract on the Blockchain, then he/she has to request the transaction. In the next step, the requested transaction is broadcasted to a Peer-to-Peer network containing computing devices called as network nodes. After the transaction and user's position is validated by the P2P network, the transaction is combined with other same date transactions for formation of a new block of data for the digital ledger. The newly prepared block is then added to the existing Blockchain wherein the transaction is permanent and non-modifiable. After this stage, the transaction is completed. The fact is that cryptocurrency has no natural value and it cannot be used for other services. Also cryptocurrency does not have physical form and has its existence only in that network. No third party centralized authority or banking institution is there and it is totally decentralized.

17.3 Blockchain Systems and Cryptocurrencies

The requirement of the industries such as healthcare, real estate, education, politics, etc. is different. Blockchain has to evolve according to the application for which it is being used. Blockchain is a multi-facet technology. There are different types of Blockchain from fully open, permissionless (public) to permissioned (private), depending on various features and covering large array of systems.

17.3.1 Public (Permissionless) Blockchain

A user can join or leave the network as per his wish in the permissionless open public Blockchain. Here the consent from any central entity is not needed. Only user needs to join the network and add transactions to the digital record keeping register which is generally a computing system wherein the necessary software is installed. The redundancy feature of public Blockchain offers security to it. Electric

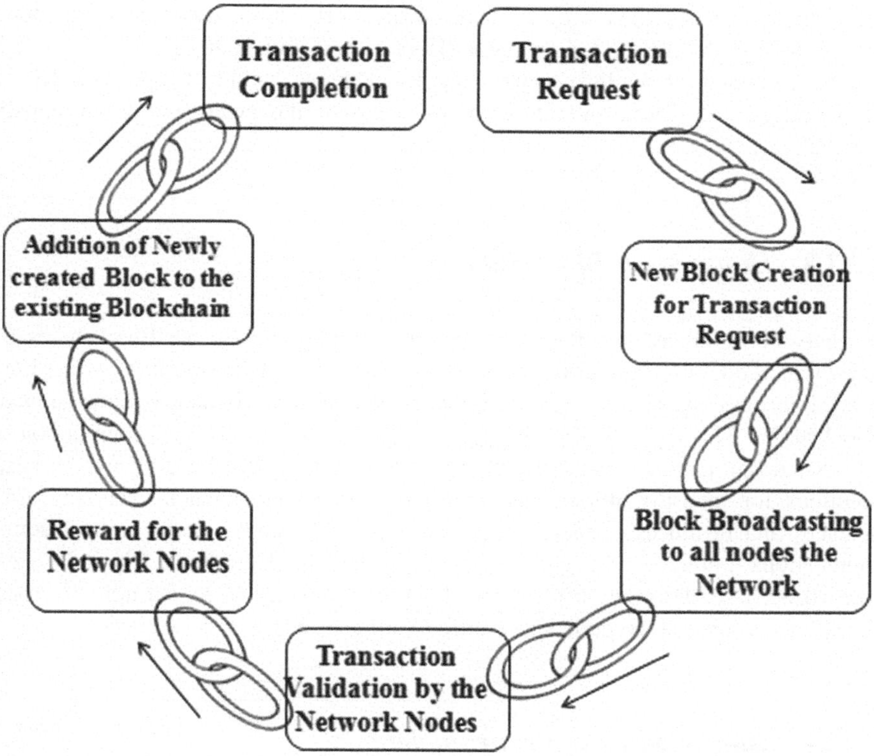

Fig. 17.2 Cryptocurrency transaction in a Blockchain

power consumption is more because of large number of network nodes and slow speed of operation. The huge benefit from public Blockchain is that it is fully transparent type of ledger with user secrecy.

Large numbers of cryptocurrencies belong to this class of public Blockchain. For example, Bitcoin, Bitcoin Cash, Litecoin.

17.3.2 Private (Permissioned) Blockchain

The rules for the ledger are pre-set by the network administrators in the private Blockchain. Nodes need to take permission from the network administrator for joining the network and initiating a transaction. The identity of the network nodes is verified before they join the network. There are two subclasses of private Blockchains such as public permissioned Blockchains and enterprise permissioned Blockchains. In public permissioned Blockchains, anyone can view or access the Blockchain but only authorized network nodes have permission to initiate the new transactions or update the state of the ledger. In case of enterprise permissioned

Blockchains, the access is restricted and only network administrator can create new transactions or update the ledger status (Houben and Snyers 2018).

The examples of public permissioned Blockchains include Ripple and NEO. Hyperledger, R3 Corda and Quorum are some of the enterprise permissioned Blockchains.

17.3.3 Consortium Blockchain

Partially decentralized and the combination of public and private Blockchains is known as the Consortium Blockchain. It is a hybrid combination in between low trusted permissionless and a single highly trusted authority model of permissioned Blockchains. This type of Blockchains generally functions under the group of board of managing directors or higher authorities. Consortium Blockchains are the best solutions for inter-discipline, cross-industry applications, financial services, life sciences and health care, energy and resources, technology, media and telecommunications, public sector, and consumer and industrial products. Big industries like IBM, SAP, Daimler, Fijitsu, and Intel have formed a consortium of cross disciplines known as Hyperledger Fabric (Yafimava 2019).

17.3.4 Blockchain Cryptocurrencies

Blockchain is known to be a digitally distributed record register which cannot be altered. Blockchain is considered to be the strength of cryptocurrency market and it is basically the technology running behind variety of cryptocurrencies. Some of the famous Blockchain cryptocurrencies are listed in Fig. 17.3. Cryptocurrency user is a person who is an owner of the digital currency coins and can use these to buy real or virtual goods and/or services can make payments for P2P transactions, or the user can keep these coins as an asset.

Miner is a person who can solve the cryptographic puzzle to participate in the transaction validation process that is based on consensus system. Cryptocurrency users are offered exchange services by cryptocurrency exchanges by incurring some costs. They allow users to buy or sell their coins for fiat currency. Some of the renowned cryptocurrencies are Bitfinex, HitBTC, Kraken, and Coinbase GDAX. Trading platforms are also available for the users who want to buy or sell coins with cash. The example is eBay for cryptocurrencies. Different wallet providers, coin inventors, and coin offerors are also available (Houben and Snyers 2018).

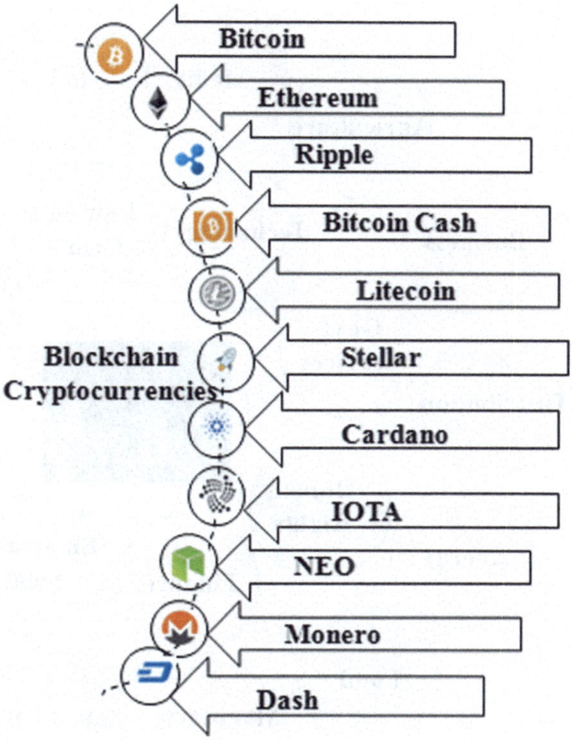

Fig. 17.3 Various Blockchain cryptocurrencies

17.4 Applications of Blockchain

For the purpose of overcoming some of the flaws in the existing banking system, Blockchain was developed. Instead of paying transaction processing fees to the banking institutions working as a third party entity, Blockchain skipped the third party validation and it is done by the network nodes by themselves. There are numerous applications of Blockchain from fund transfer to food and manufacturing industry till Government policy sectors as illustrated in Fig. 17.4.

For money transfer and payment processing, in between banking entities are avoided in the Blockchain that can increase the speed of banking operations by manifold. Supply chain monitoring by both the business persons as well as customers can be done efficiently with the Blockchain because of its open structure. This is very important aspect for assurance of quality control. Loyalty rewards can be provided in the retail industry with the help of Blockchain. Some kind of incentives can be given to customers for growth of the business.

Copyright infringement can be effectively implemented and monitored with the help of Blockchain techniques. Voting scams can be eliminated with framing digital voting using Blockchain wherein the filled in information cannot be altered. For legal ownerships, Blockchain can prove a milestone for real estate, land and other

Fig. 17.4 Applications of Blockchain

title transfer applications. Food chain can be tracked with Blockchain from source to destination with great ease. For tax regulation and compliance, Blockchain can prove a strong resource for recording the sales and other necessary information and to keep it intact (Williams 2018).

Digital record keeping in healthcare industry is of prime importance that can protect patient privacy. Internet of Things (IoT) can be effortlessly managed with Blockchain by maintaining trust on the network in use. Interoperability and secure data transactions are the important requirements of IoT which can be effectively taken care by Blockchain. Trading in the energy sector can become considerably faster with the Blockchain usage in efficient utilization of the resources and good regulatory fulfillment.

Agricultural applications including soil data records processing, agro products business, sales and marketing task, crop yields data, etc. can be proficiently monitored and stored with the help of Blockchain with due security and privacy. In the energy sector, raw material data, energy generation data, availability of resources, energy supply and demand balance statistics, tariff data maintenance, utility condition monitoring are very critical jobs that can be monitored and managed effectively by Blockchain. Smart City is a buzz word today. Various service offerings, energy and water management data, pollution control data, digital transactions awareness and encouragement for that, etc. are the necessary applications that are expected to be handled by Blockchain technology (Kumar and Mallick 2018).

17.5 Threats and Attacks

Weaknesses in the Bitcoin currency can ultimately result into the potential threats to the Blockchain technology and its significant applications. Blockchain is very secure and immutable to the attacks by default. If thousands and tons of addresses are generated through the database of Blockchain public database, then also it is not a threat because nobody can modify that data. Small issues like breaking cryptographic algorithms, scalability problems, attacking all the users who are part of Blockchain, Transaction spamming, can be easily handled by Blockchain structure itself. The real threats are if digital wallet is exposed to threat, and if someone is successful in tracing the coin's history, in case of Sybil attack, packet sniffing, DoS attack, and if some malicious users are trying to insert illegitimate contents on the Blockchain. Various attacks are evolving related to Blockchain as shown in Fig. 17.5 that include selfish mining attack, stubborn mining attack, eclipse attack, balance attack, Sybil attack and stalker attack.

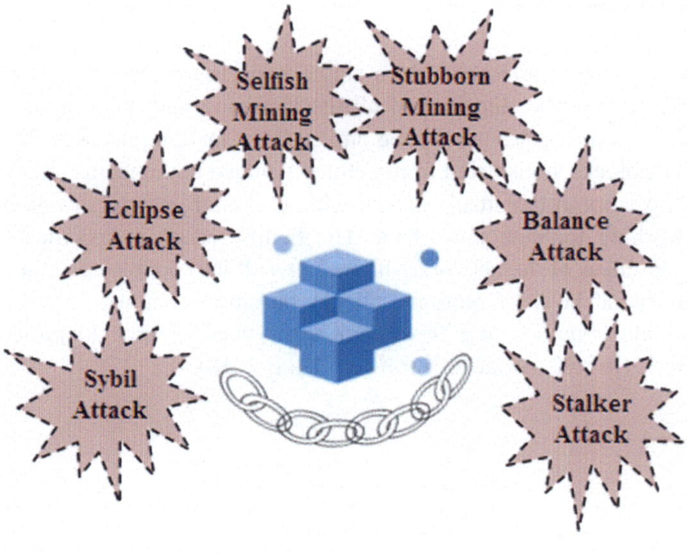

Fig. 17.5 Attacks on Blockchain

17.5.1 Selfish Mining Attack

A chain is created in secret and released by the miners who want maximize their profit. Actually this strategy of optimizing the mining for malign benefits, offends the other legitimate miners and their authorities. This type of attack is also known as block withholding attack because already verified mined block is held by the selfish miners from broadcasting it to other miners. Once the block is withheld form broadcasting in the group, selfish miner shows more proof of work and tries to get rewards. For large parameter space, selfish mining is found to be not optimal.

17.5.2 Stubborn Mining Attack

With more stubborn strategies, the selfish miner tries to get more revenues. Stubborn mining is found to be more beneficial if the attack space is large. There are three main types of Stubborn Miners such as lead stubborn, trail stubborn, and equal fork stubborn. The authors in (Nayak et al. 2016) used Markov chain demonstration to additionally exploit network level attacks to increase the miner's revenue. A block is immediately published if it is mined by a legitimate node but if it is mined by malign actor, it tries to hide it. Then attacker tries to gather other information such as total blocks in each chain, leading chain, and number of legitimate or malicious nodes in the chain.

17.5.3 Eclipse Attack

Instead of targeting whole network, a single node is targeted in case of Eclipse attack. Then the node is deprived of all the incoming and outgoing communications. And all the messages are filtered and used for malign purposes. The targeted victim and its blocks are isolated from getting included in the mainstream chain and it is forced to work in the attacker chain without the target being aware of the fact that it is working for malicious chain. The Eclipse attack is specifically used to crack the consensus of the network which can result into double spending problem, attacks on second layer of protocol, attacks on smart contracts, etc. Cross layer design to build complicated P2P system is the need of time to protect against Eclipse attack (IoTeX in HackerNoon.com Blog 2018).

17.5.4 The Balance Attack

This attack mainly tries to delay the network communications between multiple node subgroups along with taking care of balancing of mining power. The tradeoff is found in between network delay and mining power. The Balance attack powers the Ghost protocol which works mainly on sibling or uncle blocks for block selection purpose. With the help of these blocks, the attacker can successfully mine a single branch and rest of the network is untouched. This mined branch is merged into the competing Blockchain to control the branch selection process. The miners are divided into two equal groups and the communication delay in between these two groups is increased (Natoli and Gramoli 2016).

17.5.5 The Sybil Attack

Through this attack, a malign intent person can run multiple nodes on a Blockchain network. These nodes create Sybil identities and can out-vote the legitimate nodes from the consensus mechanism. The Sybil attack nodes can change the organization of the deals and can stop transactions being finalized. These nodes can reverse the business deals that make the system to double spending problem. Behind the scenes, a single malign intent entity handles all these forged identities in the network. Blockchain uses different consensus algorithms to combat against Sybil attacks including proof of work, proof of stake, and delegated proof of stake. For the creation of new blocks, Blockchain applies explicit set of protocols. The incentives are there for the honest mining nodes (Garner 2018).

17.5.6 The Stalker Mining Attack

The stalker mining attack is a modified version of selfish mining attack. In selfish mining attack, the attacker tries to increase the gains whereas in the stalker mining attack, the attacked node tries to deny for the service without thinking of gain. Two important strategies are applied by the stalker miner that consist of decision to choose the best moment for disclose the private chain and the moment to decide to accept the public chain. The stalker miner uses adopt, wait, and publish heuristic algorithms for this attack. For restarting the attack, the invader takes on the honest chain. Without revealing its identity, the attacker digs for sensitive information. When the attacker chain becomes longer than the honest chain, and when sufferer publishes a new block, then the attacker's chain is disclosed (Jesus et al. 2018).

17.6 Cyber Security Revolutions with Blockchain

Cyber-attacks are growing day by day. Attackers are applying new vectors and on daily basis, thousands of cyber-attacks are launched. The users and industries are facing lot of challenges in keeping their sensitive information safe. No industry or customer is safe in the world of various wireless technologies. Blockchain technology has emerged as a ray of hope for providing effective cyber security against intelligent attackers. Blockchain has capability to develop robust security against the shrewd cyber-attacks. Figure 17.6 illustrates possibility of various cyber security revolutions with the help of Blockchain technology.

User awareness and preparedness is the most important part of the success of any security solution. Viruses, worms, and Trojan horses enter the user's computing system through various software downloads or mobile device application updates. Blockchain can allot unique hashes to this kind of downloads and updates. The user can cross check these hash values and can refrain from deceitful viruses and malwares. Verification of software downloads and mobile application updates is a first step towards the cyber security.

Distributed denial of service attacks are targeting sophisticated business entities like financial institutions, movie buildings, industries, etc. The DDoS attack frequency has increased by manifold. Due to growth of IoT penetration, the possibility of DDoS attacks has been increased by huge amount. With the cognitive capability, Blockchain technology can reduce the percentage of these attacks. Users can give their spectrum on lease for other users thereby helping to serve the surplus traffic in the networks.

Biometric techniques at the place of passwords can be the good alternative to weak passwords. User accounts on the social networking sites have been hacked due to the weak passwords. Blockchain does not make use of passwords. It uses biometric mechanisms, private keys and multi-step authentication to make more secure and strong. Automation systems are more prone to cyber attacks. Blockchain technology and its advances can identify fraudulent commands and inputs. After identification of such sources, these can be kept at a bay with other security techniques.

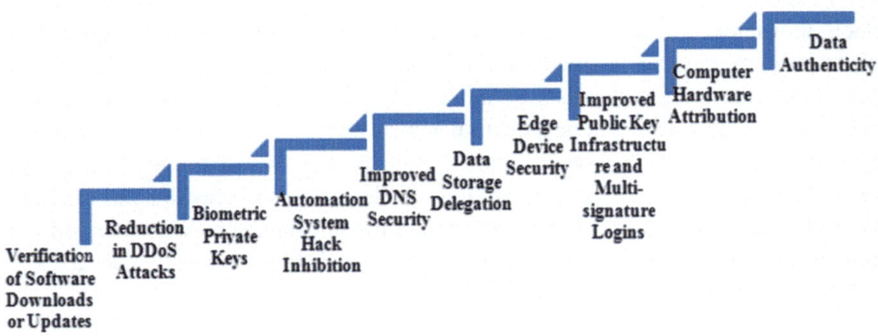

Fig. 17.6 Cyber security revolutions with Blockchain

Domain name system that is DNS can be made safer with the Blockchain techniques with its decentralized structure. Centralized systems can have more harm from DDoS attacks as compared to decentralized systems because of distributed point of network entities. Huge data is generated in this wireless mobile application age. The data storage with centralized locations is a risky affair. Single point of contact is always more prone to security attacks. Because of its distributed nature, Blockchain is becoming effective solution for many of the security threats. Today's communication systems are the converged version of various applications, Internet, and other technologies. These technologies have become integral part of human life. The edge devices are vulnerable to various threats. Blockchain's decentralized architecture and multi-level authentication is expected to provide better security.

Multi-signature authentication model is the basic building block of Blockchain technology. Secure access to network with the help of Blockchain can reduce the hack frequency, and can protect valuable user data. Computer hardware attribution and prevention of strange intrusions can be obtained through Blockchain techniques. Its immutable digital record ledger can save the information related to manufacturing and transport details of the computing node in the system. False data injection into the legitimate system or digital ledger can be prevented through the use of Blockchains. False positives in any cyber system always wastes money, energy and time. Of The kind of bad data is filtered at every step in the Blockchain thereby reducing the chance of entry into the system (Mire 2018).

17.7 Summary

Last year various government agencies, universities, power companies and well known industries have faced massive cyber-attacks. Due to these attacks, these entities have set cyber security as the highest priority. Blockchain technology based solutions are the most adopted practice by the companies and users. For prevention of DDoS attacks and other IoT enabled threats, distributed and immutable Blockchain techniques are proving to be eye catching techniques. Security is not a onetime solution; it is a continuously evolving process. With variety of vectors and techniques used by hackers, Blockchain technology and other parallel cyber security solutions need to grow more maturely and strongly.

References

Garner B (2018) What's a sybil attack & how do Blockchains mitigate them? COINCENTRAL Blogs. https://coincentral.com/sybil-attack-blockchain/

Houben R, Snyers A (2018) Cryptocurrencies and Blockchain—legal context and implications for financial crime, money laundering and tax evasion. In: European Parliament Study Requested by TAX3 committee

IoTeX in HackerNoon.com Blog (2018) Eclipse attacks on Blockchains' peer-to-peer network. https://hackernoon.com/eclipse-attacks-on-blockchains-peer-to-peer-network-26a62f85f11

Jesus EF, Chicarino VRL, de Albuquerque CVN, Rocha AADA (2018) A survey of how to use Blockchain to secure internet of things and the stalker attack. Hindawi J Secur Commun Netw 2018:1–28. Article ID 9675050

Kumar NM, Mallick PK (2018) Blockchain technology for security issues and challenges in IoT. In: Proceedings of Elsevier science direct international conference on computational intelligence and data science (ICCIDS), pp 1815–1823

Mire S (2018) Disruptor daily blog on, "Blockchain in cybersecurity: 10 possible use cases. https://www.disruptordaily.com/blockchain-use-cases-cyber-security/

Nakamoto S (2008) Bitcoin: a peer-to-peer electronic cash system. Bitcoin White Paper

Natoli C, Gramoli V (2016) The balance attack against proof-of-work Blockchains: the R3 testbed as an example. ArXiv J 1612

Nayak K, Kumar S, Miller A, Shi E (2016) Stubborn mining: generalizing selfish mining and combining with an eclipse attack. In: IEEE European symposium on security and privacy (EuroS&P)

Priyadarshini I (2019) Introduction to Blockchain technology. In: Le D-N et al (eds) Cyber security in parallel and distributed computing, pp 93–262

Pro Blockchain Media Blog (2018) The emergence of Blockchain 5.0. https://pro-blockchain.com/en/the-emergence-of-blockchain-5-0

Unibright Blog (2017) Blockchain evolution: from 1.0 to 4.0. https://medium.com/@UnibrightIO/blockchain-evolution-from-1-0-to-4-0-3fbdbccfc666

Williams S (2018) The motley fool blog on "20 real-world uses for Blockchain technology". https://www.fool.com/investing/2018/04/11/20-real-world-uses-for-blockchain-technology.aspx

Yafimava D (2019) Openledger insights blog on "What are consortium Blockchains, and what purpose do they serve?". https://openledger.info/insights/consortium-blockchains/

Chapter 18
Research Challenges and Future Scope

The cyber security has several potential open research challenges as well as future scope. There are various flaws in the ICT and cyber systems such as level code defects, trustworthy computing and selfish behavior of the users and machines, security architecture imperfections, privacy shortcomings, usability issues, and weak security metrics. To overcome these flaws with effective solutions is a challenging task. Key research challenges and future scope is discussed subject wise in this chapter.

18.1 Cyber Threats and Attack Overview

As the whole world is looking forward to next generation of converged wired and wireless networks, numbers of new vulnerabilities are emerging. These include spyware, virus, malware, phishing, hacking, spam, Internet worm, Hacktivist, Identity thefts, password scams, firewall vulnerabilities, etc. To combat these cyber threats and attacks, design and development of usable security solutions is very much essential (Neumann 2000; Liska 2003; Holkar et al. 2013; FOSSBYTES 2017).

18.2 Phishing

Users should always be prepared to handle the incidents. Handling the incidences, it is very crucial to maintain reliability, privacy and security of the data and persons involved in those incidences. To observe the post incidence situation, to analyze the circumstances, to track the people and data for the final conclusions and ultimately to present it accurately is a difficult task. There is a need for further research on all

R. Prasad and V. Rohokale, *Cyber Security: The Lifeline of Information and Communication Technology*, Springer Series in Wireless Technology,
https://doi.org/10.1007/978-3-030-31703-4_18

these crucial issues which can build more strong mechanisms for efficient and quick incident handling responses.

Collaborative information sharing should be critically secure, reliable and privacy preserved. Enterprises or institutions can reduce the rate of incidents by actually securing networks, computing systems and applications used. Prioritization of the incidents and well planning for handling such incidents is the essential thing. Lessons learned during such circumstances prove to be very useful for future planning (Bisson 2016; Lord 2018).

18.3 BOTNETs

1. The security systems which are presently available focus on just one network, i.e., either HTTP or TCP but not both. Hence, detection algorithm can be researched for both the networks.
2. Using the world wide database, one can have a strong detection of BOTNETS. Also, a system can be implemented where the detection setup is in place before LAN services reach the user. This may help in protecting the computer from getting affected when connected to LAN.
3. From a state of the art, it has been understood that real-time detection has not yet focused much. To have real time detection, the system must be developed which can also identify new attacks rather than just identifying with the help of stored database.
4. Many systems developed till now are protocol dependent. Hence, if the memorandum of botnet communication is changed, it becomes difficult to identify attacks. Therefore, future study can be more focused on real time protocol independent detection of botnets.
5. Cryptography is one of the safest methods to avoid illegitimate attacks. With the help of strong cryptographic security solutions, a communication channel can be protected within the botnets, and also, it can ensure authenticity and integrity of botmaster commands.
6. Botnet development system must be implemented in such a way that advance bot designs can also be identified (Geers 2011; Rodrıguez Gomez et al. 2013).
7. Research work must have more focus on authentication of legitimate system users which can help in filtering out attackers.
8. There should be collaboration and cooperation among the key defense points among service providers throughout The Internet.

18.4 Malware

Malware may be propagated using spam, and may also be used to send spam. It may take advantage of bugs. Malware may be used to mount denials of service attacks. Addressing the problem of malware is vital for improving computer security. Malware has connection with many other attacks. Because of this, it is really challenging for the researchers to find new security solutions which will be secure as well as usable (Stephen and Lee 2014; Liţă et al. 2017).

18.5 Copyright Infringement

Watermarking is a process in which additional data is embedded along with the underlying data i.e. audio, images, and video. In future, the watermark can be equipped with intelligence that can potentially tell the content of an audio file, where it originated from, the distribution channel, date distributed, to whom it was distributed (legal customer profile) and possibly the type of media on which the track was last used on. Also in future, research work may be extended for reduction of the capacity for audio watermarking with more effectiveness.

There are various technical challenges in watermarking research. The robustness and imperceptibility trade-off makes the study quite interesting. To attain imperceptibility, the watermark should be added to the high-frequency components of the original signal. On the other hand, for robustness, the watermark can be added to the low-frequency components only. Thus, the watermarking scheme can be successful if the low-frequency elements of the original signal are used as the host for watermark insertion (El-Wahed et al. 2007; Bhargava et al. 2012).

18.6 Cyber Forensics

The cyber forensics investigations and finding out the evidences demand completeness as well as speed. It has become imperative for today's world to design and develop the quick and perfect digital forensics analysis mechanisms. Digital forensics is in reality the reverse engineering task with lot of challenges such as abstraction and modularization. Also standards are needed to be developed for data and code interchange mechanisms. There are various research challenges for cyber forensics field including data diversity, device variety, network mixtures, converged services, dispersed evidence, big volumes of evidence data, trust issues related to audit data, testing and validation diversities and issues, and anti-forensic activities. There is need of uniform policies for all these problems and related investigations on the global front (Choi et al. 2016; Saibharath and Geethakumari 2015).

18.7 Cloud Computing

Trustworthy computing, privacy preservation, Big Data and usable security solutions are the critical challenges for cloud computing. Achievement of effective fault tolerance mechanism for improvement in serviceability, reliability in cloud computing environment and design of a foolproof fault tolerance strategy on voluntary cloud infrastructure is the need of current cloud computing system. Fault tolerance as a service in cloud computing requires a deep analysis and consideration of complexity, inter-dependability, and occurrence of multiple instances of an application running on several virtual machines. This is a worthy research direction (Halabi et al. 2018; Puri and Agnihotri 2017).

18.8 IoT and M2M

IoT involves tagging, sensing, minimizing, and thinking capabilities to be embedded into the objects surrounding us. The user wants smartness in the objects or devices surrounding them. Intelligence comes with the abilities of decision making, controlling the situation and commanding the circumstances. For IoT, data transparency, data integrity and anonymization of data are the real challenges in front of the researchers.

M2M Communications ecosystem needs standardization, fragmentation, complexity handling capabilities and ultimately, better user experience. On device level, size reduction, energy consumption reduction, less memory and computational capabilities are the central issues for research purpose. Data Confidentiality, integrity, and availability are the crucial questions. Some regular routines for vulnerability scanning and monitoring are needed to be developed. Advanced security technology standards are to be primarily designed. For better risk management capabilities in the machines or devices, security policy enhancements are the must. Advances are necessary for the regular monitoring of some specific cyber threat profiles. Usable security with privacy and reliability are the main requirements for sustainable IoT and M2M (Hussain et al. 2016; Intelligence 2014).

18.9 Smart Grid

Smart Grid provides some services that can help utilities to transport energy more efficiently and enable consumers to handle their energy usage more efficiently. These applications and services will have certain communication requirements that cannot be satisfied by the current communication infrastructure. Therefore, the major challenge that needs to be addressed the changes in modern infrastructure to meet these necessities and facilitate the recognition of the AMI. The main challenges that should be considered while designing the infrastructure of AMI are privacy preservation, secure data collection and secure data storage and reducing processing cost (Xiang and Wang 2019; Buyuk 2018).

18.10 Bluetooth Communication

Usage of Bluetooth in smart phones and laptops has increased enormously and is growing tremendously day by day in all types of surroundings. However, Bluetooth technology has not entirely caring security issues in the standardization process and is more vulnerable to different kind of attacks as compared to other technologies. With the help of many algorithms like IDEA, AES, MD5 and RSA, researchers can implement many hybrid algorithms for enhancement of security in Bluetooth communication. Different wireless technologies are available in these devices, but along with that, new ways or mechanisms to secure communications must be developed keeping them simple, efficient, functional and practical. The future research work may make use of Galois field for enhancing security of communication with the use of Bluetooth technology (Ben Henda 2014; CYWARE malware and Vulnerabilities Blog 2019).

18.11 E-commerce

E-commerce security is the protection of E-commerce assets from unauthorized access, use, alteration, or destruction. The main safety issues of E-commerce traditional authentication mechanism based on identity to provide security or access control; also, traditional encryption and authentication algorithms require high computing power of computer equipment. Online payment systems of the world are maturing day by day. For a full-fledged E-commerce infrastructure and system of the world to operate smoothly, we need a strong and secure online payment mechanism. Although an integrated banking system is in the pipeline yet the present Internet banking system, online payments mechanism, etc., are not up to the mark. Further, the modern banking, financial and regulatory environment of developing countries need to be streamlined. Online shopping and E-commerce in developing countries must be encouraged but at the same time, legal and cyber security issues must also be taken seriously.

Trojan horse programs launched against client systems pose the greatest threat to e-commerce because they can bypass or subvert most of the authentication and authorization mechanisms used in an e-commerce transaction. These programs can be installed on a remote computer by the simplest of means: email attachments. Cyber security uses different prediction algorithms which are considered for predicting cyber-attacks. Though the prediction techniques being applied at present in the field of cyber security seem to be capable of enhancing cyber security measures, still require a lot of research to deal with an influx of new threats in large wireless computer networks. In future, a strong security mechanism is needed which can secure the e-commerce transactions from all type of threats and attacks (Niranjana Murthy and Chahar 2013; Farshchi et al. 2011).

18.12 Social Networking

With the increased number of threats to networks such as worms, viruses and DOS, security can no longer be viewed as an option, even within "private" networks. Securing all equipment, resources, and networks these days, is very critical to maintain uptime and seamless access to services. Historically, this has been the largest barrier to the implementations of the security system and policies. Today, the amount of time spent in repairing a network due to just a single worm or virus attack can easily be greater than the exact time required to secure more adequately an enterprise. Defending against the session hijack attack is tough because the attack is not dependent on software vulnerabilities, but rather, protocol limitations within the TCP/IP protocol. Fortunately, there are many options in systems and software to increase the security of the network while reducing the overhead of managing such systems. Even basic practices such as periodic software updates, locking down all devices and using centralized authentication and secure access methods can go a long way in reducing risks. A variety of robust security techniques is needed which can reduce user's exposure to the attacks including intrusion detection and intrusion prevention systems, firewall configuration, and IPsec, etc (Asian Schools of cyber law 2010; Ghosh 2019).

18.13 Secure Incident Handling

Aftermath strategy once the attack incident takes place is of prime importance. For every organization it is must to build a responsible incident response team (IRT). Preparedness of incident response team should be regularly challenged and cross checked. For the normal functioning of the incident response team, some protocols should be designed and maintained. Effectiveness of IRT must be monitored with some operational metrics. The positive as well as negative results should be informed to the team members through friendly communication. Involvement of multi-disciplinary areas and persons is necessary. Considerable budget should be allocated for cyber security measures as well as incident response. The organization should come ahead in discussing threat indicators to third party organizations with future collaboration perspective (The Government of the Hong Kong Special Administrative Region 2012; United States Government Accountability Office 2014).

18.14 Smart Devices

In today's mobile world, the user makes use of his smart phone for almost every task including email accessing to bank transactions. The smart phone has become digital wallet for the users. Since the users are utilizing their smart phones for sensitive tasks such as online banking and online payments, they have become the obvious and money-spinning targets for the cyber criminals. This poses very big challenge for security experts. Device data exfiltration, data tampering, data or device loss and security from advanced malware are the tough challenges for the modern smart devices (Raphael 2019; Tekade and Shelke 2014).

18.15 AI and ML in Cyber Security

AI is adding value to the security sectors of the corporations and individual persons as well, it has helped a lot in eliminating cyber-attacks to a great extent resulting in securing the data for companies and avoiding many frauds that could happen. With the added benefit of machine learning, AI algorithms can easily turn down the cyber security breaches and with the proper information, the security personnel can rectify the system within fraction of time. So, it is evident that the artificial intelligence plays a very important role in cyber security. There is huge scope for upcoming researchers to work in this field (Tyugu 2011; Garbade 2019).

18.16 Blockchain Technology

Blockchain technology based solutions are the most adopted practice by the companies and users. For prevention of DDoS attacks and other IoT enabled threats, distributed and immutable Blockchain techniques are proving to be eye catching techniques. Security is not a onetime solution; it is a continuously evolving process. With variety of vectors and techniques used by hackers, Blockchain technology and other parallel cyber security solutions need to grow more maturely and strongly (Priyadarshini 2019; Houben and Snyers 2018).

18.17 Summary

For the converged next-generation networks, it is very much essential to provide self-healing nature with aggressive and self-protective characteristics capabilities. Also, there is the need for the formation of international laws so that uniform for cyber security policies will be worldwide uniform. Global Internet governance model should be developed with high security and privacy provisions. Usable and trustworthy security solutions are the most urgent needs of the cyber world.

References

Asian Schools of cyber law (2010) A to Z of cybercrimes. Department of Science and Technology Government of India, Lexcode Education and Assessment Platform (LEAP)

Ben Henda N (2014) Generic and efficient attacker models in SPIN. In: SPIN '14, San Jose, USA, 21–23 July 2014

Bhargava N, Sharma MM, Garhwal AS, Mathuria M (2012) Digital image authentication system based on digital watermarking. In: International conference on radar, communication and computing (ICRCC), pp 185–189

Bisson D (2016) 6 common phishing attacks and how to protect against them. Tripwire-the state of security. https://www.tripwire.com/state-of-security/security-awareness/6-common-phishing-attacks-and-how-to-protect-against-them/

Buyuk OO (2018) A novel actual time cyber security approach to smart grids. In: 6th international Istanbul smart grids and cities congress and fair (ICSG)

Choi Y, Lee J-Y, Choi S, Kim J-H, Kim I (2016) Introduction to a network forensics system for cyber incidents analysis. In: 18th international conference on advanced communication technology (ICACT)

CYWARE malware and Vulnerabilities Blog (2019) Latest Bluetooth hacking techniques expose new attack vectors for hackers. https://cyware.com/news/latest-bluetooth-hacking-techniques-expose-new-attack-vectors-for-hackers-a16cfb5e

El-Wahed SA, Elfatatry A, Abougabal MS (2007) A new look at software plagiarism investigation and copyright infringement. In: ITI 5th international conference on information and communications technology, pp 315–318

Farshchi SMR, Gharib F, Ziyaee R (2011) Study of security issues on traditional and new generation of e-commerce model. In: International conference on software and computer applications

FOSSBYTES (2017 Feb 28) What is social engineering? What are different types of social engineering attacks? https://fossbytes.com/what-is-social-engineering-types-techniques/

Garbade MJ (2019) Top 8 open source AI technologies in machine learning. opensource.com. https://opensource.com/article/18/5/top-8-open-source-ai-technologies-machine-learning

Geers K (2011) Strategic cyber security. CCDCOE, NATO Cooperative Cyber Defense Center of Excellence, 19–22 June 2011

Ghosh S (2019) Top seven social media threats. Computer Weekly Blog. https://www.computerweekly.com/tip/Top-seven-social-media-threats

Halabi T, Bellaiche M, Abusitta A (2018) A cooperative game for online cloud federation formation based on security risk assessment. In: 5th IEEE international conference on cyber security and cloud computing (CSCloud) and 4th IEEE international conference on edge computing and scalable cloud (EdgeCom), pp 83–88

Holkar AM, Holkar NS, Nitnawwre D (2013) Investigative analysis of repudiation attack on MANET with different routing protocols. Int J Emerg Trends Technol Comput Sci (IJETTCS) 2(3)

Houben R, Snyers A (2018) Cryptocurrencies and Blockchain—legal context and implications for financial crime, money laundering and tax evasion. In: European Parliament Study Requested by TAX3 committee

Hussain F, Ferdouse L, Karim L (2016) Security threats in M2M networks: a survey with case study. Comput Syst Sci Eng

GSMA Intelligence (2014 Feb) Analysis from concept to delivery: the M2M market today
Liska A (2003) Network security: understanding types of attacks. Pearson InformIT. http://www.informit.com/articles/article.aspx?p=31964&seqNum=3
Liţă CV, Cosovan D, Gavriluţ D (2017) Anti-emulation trends in modern packers: a survey on the evolution of anti-emulation techniques in UPA packers. J Comput Virol Hacking Tech
Lord N (2018) What is a phishing attack? Defining and identifying different types of phishing attacks. Digital Guardian Blog. https://digitalguardian.com/blog/what-phishing-attack-defining-and-identifying-different-types-phishing-attacks
Neumann PG (2000) Denial-of-service attacks. In: Communications of the ACM, p 136. Academic OneFile
Niranjana Murthy M, Chahar D (2013) The study of e-commerce security issues and solutions. Int J Adv Res Comput Commun Eng 2(7)
Priyadarshini I (2019) Introduction to blockchain technology. In: Le D-N et al (eds) Cyber security in parallel and distributed computing, pp 93–262
Puri S, Agnihotri M (2017) A proactive approach for cyber attack mitigation in cloud network. In: International conference on energy, communication, data analytics and soft computing (ICECDS)
Raphael JR (2019) 7 mobile security threats you should take seriously in 2019. https://www.csoonline.com/article/3241727/7-mobile-security-threats-you-should-take-seriously-in-2019.html
Rodrıguez Gomez RA, Macia Fernandez G, García-Teodoro P (2013) Survey and taxonomy of botnet research through life-cycle. ACM Comput Surv 4–10:45. Article 45
Saibharath S, Geethakumari G (2015) Cloud forensics: evidence collection and preliminary analysis. In: IEEE international advance computing conference (IACC)
Stephen C, Lee A (2014) Malware is called malicious for a reason: the risks of weaponizing code. In: 6th international conference on cyber conflict
Tekade PS, Shelke CJ (2014) A survey on different attacks on mobile devices and its security. Int J Appl Innovation Eng Manage (IJAIEM) 3(2). ISSN 2319 - 4847
The Government of the Hong Kong Special Administrative Region (2012 Sept) Information security incident handling guidelines. Version: 5.0
Tyugu E (2011) Artificial intelligence in cyber defense. In: Czosseck C, Tyugu E, Wingfield T (eds) 2011 3rd international conference on cyber conflict, Tallinn, Estonia
United States Government Accountability Office (2014 Apr) Information security-agencies need to improve cyber incident response practices. Report to Congressional Requesters
Xiang Y, Wang L (2019) An improved defender-attacker-defender model for transmission line defense considering offensive resource uncertainties. IEEE Trans Smart Grid 10(3):2534–2546

Index

R. Prasad and V. Rohokale, *Cyber Security: The Lifeline of Information and Communication Technology*, Springer Series in Wireless Technology,
https://doi.org/10.1007/978-3-030-31703-4

The manufacturer's authorised representative in the EU is Springer Nature Customer Service Centre GmbH, Europaplatz 3, 69115 Heidelberg, Germany. If you have any concerns regarding our products, please contact ProductSafety@springernature.com

Printed and bound by CPI Group (UK) Ltd, Croydon, CR0 4YY
25/09/2026
02222892-0010